Principles of Automation and Control

Edited By

Ilesanmi Afolabi Daniyan
Department of Industrial Engineering
Tshwane University of Technology
Pretoria-0001
South Africa

Principles of Automation and Control

Editor: Ilesanmi Afolabi Daniyan

ISBN (Online): 978-981-5080-92-6

ISBN (Print): 978-981-5080-93-3

ISBN (Paperback): 978-981-5080-94-0

Published by Bentham Science Publishers Pte. Ltd. Singapore. All Rights Reserved.

First published in 2023.

need for a court order if at any point you breach any terms of this License Agreement. In no event will any delay or failure by Bentham Science Publishers in enforcing your compliance with this License Agreement constitute a waiver of any of its rights.

3. You acknowledge that you have read this License Agreement, and agree to be bound by its terms and conditions. To the extent that any other terms and conditions presented on any website of Bentham Science Publishers conflict with, or are inconsistent with, the terms and conditions set out in this License Agreement, you acknowledge that the terms and conditions set out in this License Agreement shall prevail.

Bentham Science Publishers Pte. Ltd.
80 Robinson Road #02-00
Singapore 068898
Singapore
Email: subscriptions@benthamscience.net

BENTHAM SCIENCE

CONTENTS

FOREWORD

The dynamics of consumer needs, as well as the need to ensure time and cost effectiveness of the manufacturing process without compromise to product's quality necessitate the development of robust automation and control strategies. With the advent of Industry 4.0, industries are looking for ways to achieve sustainability, cost efficiency, competitive advantages, productivity and profitability, economies of scale, scalability and flexibility, hence, the need for the automation of industrial systems and processes. The automation and control of industrial systems focus on the efficiency of industrial processes and organisation's profitability. This will also improve system's flexibility with a substantial improvement in the manufacturing processes, product quality, and a reduction in industrial errors.

This book presents recent and novel theoretical concepts and practical findings in the field of automation and control. It comprises thirteen chapters that delve into the principles of automation and control. Furthermore the following thematic areas were also covered: automated processes and systems, control theory, system's control, computer control devices, industrial automation tools, application of industrial automation as well as practical examples of how automation can be achieved in systems. Furthermore, empirical findings that highlight recent advances in the field of automation and control are discussed.

The book also discusses the performance of automated systems and controls and how such performances can be enhanced with relevant examples and case studies drawn from real world research. The book is multi-disciplinary in nature and it is designed for learners, experts, instructors, academics, engineers, and professionals whose field of interests match the niche area of automation and control.

The book disseminates knowledge critical for system's automation and control. The book provides entrepreneurs, experts and engineers meaningful insights into practical ways to achieve automated business solutions from the design level. Thus, this book demonstrates how simple or complex processes can be automated and controlled to deliver a sustainable business value.

With the knowledge and expertise of the contributing authors, my teaching and research acumen in manufacturing, automation and control was geared to provide novel contributions to this book. Some of the automation and control concepts presented in the book are still emerging and will be of a great benefit to the readers and experts in the quest to gain wide range of knowledge in the field of automation and control.

Festus Fameso
Department of Mechanical and Automation Engineering
Tshwane University of Technology
Pretoria
South Africa

PREFACE

With the emerging technologies, the quest for competitiveness by organisations, dynamic nature of customer requirements and market demand, need to achieve product quality and organisation's bottom line; profitability in a time and cost effective manner, the role of automation and control cannot be overemphasised. There is a constant quest by organisations to achieve the development of smart systems for effective delivery, monitoring and control. Since manufacturing industries are profit-oriented and are concerned with the precision and productivity per worker of their plants, automatic systems offer the solution of high productivity and effective control without sacrificing precision and accuracy. Furthermore, the increasing complex nature of production or manufacturing systems requires engineers to have a proper understanding of the system's dynamics, behaviour and control requirements. This book bridges the gap between the theory and practice by providing and validating practical principles of automation and control. It provides a practical guided approach to help learners, professionals and organisations achieve automation and control of industrial systems. Besides it offers a robust technical data and theoretical principles relating to automation and control. The book contains twelve chapters which address many important aspects of automation and control including automation classes and principles, control theories, instrumentation, supervisory systems and robotics amongst others. It also unveils the basic and fundamental principles underlying the design of control systems.

The book is ideal for readers including learners, academics, professionals, technical person-nel's *etc.* as an educational and instructional resource across multi-disciplinary fields such as electrical engineering, chemical engineering, mechanical engineering, electronic engineering, mechatronic engineering, computer engineering, system engineering and other related fields. It will help readers gain theoretical and practical knowledge of system's control and automation so that they can be experts in the fields.

Ilesanmi Afolabi Daniyan
Department of Industrial Engineering
Tshwane University of Technology
Pretoria-0001
South Africa

ACKNOWLEDGEMENTS

I wish to appreciate God Almighty for the grace and privilege to serve as the editor of this book. I wish to thank my precious wife: Dr. (Mrs) Oluwatoyin Esther Daniyan for her invaluable support all the time. I appreciate God in the lives of my two wonderful Generals of God Almighty; Daniel and Samuel. Many thanks to all the authors who contributed their ideas and findings to this book and to all the reviewers who made time out of their busy schedules to ensure a thorough review process. My appreciation goes to Dr. Adefemi Adeodu, Dr. Boitumelo Ramatsetse and Dr. Festus Fameso for their professional contributions.

Thank you all.

DEDICATION

This book is dedicated to God Almighty: Most Glorious, Most Magnificent, Immortal, Invisible and the Only Wise God.

List of Contributors

Abdulrahman Adama   Department of Mechanical Engineering, Federal University Oye Ekit, Nigeria

Adefemi Adeodu   Department of Mechanical Engineering, University of South Africa, Florida, South Africa

Bankole Oladapo   School of Engineering and Sustainable Developmen, De Montfort University Leiceste, UK

Boitumelo Ramatsetse   Department of Mechanical & Mechatronics Engineering, University of Stellenbosch, Stellenbosch, South Africa

Cordelia Ochuole Omoyi   Department of Mechanical Engineering, University of Calabar, Calabar, Nigeria

Fawaz A. Babajide   Department of Mechanical & Mechatronic Engineering, Afe Babalola University, Ado Ekiti, Nigeria

Felix Ale   Department of Engineering & Space Systems, National Space Research & Development Agency, Abuja, Nigeria

Ilesanmi Afolabi Daniyan   Department of Industrial Engineering, Tshwane University of Technology, Pretoria 0001, South Africa

Ikenna Uchegbu   Department of Mechanical &Afe Babalola, Ado Ekiti Mechatronic Engineering, Afe Babalola, Ado Ekiti, Nigeria

Ididiong Etudor   Department of Mechanical Engineering, Afe Babalola Universit, Ado Ekiti, Nigeria

Jacobs Kelechi   Department of Mechanical & Mechatronic Engineering, Afe Babalola University of Nigeria, Ado Ekiti, Nigeria

Kazeem Aderemi Bello   Department of Mechanical Engineering, Federal University Oye Ekit, Nigeria

Khumbulani Mpofu   Department of Industrial Engineering, Tshwane University of Technology, Pretoria 0001, South Africa

Lanre Daniyan   Department of Instrumentation, Centre for Basic Space Science, University of Nigeria, Nsukka, Nigeria

Mukondeleli Grace Kana-kana   Department of Industrial Engineering, Tshwane University of Technology, Pretoria 0001, South Africa

Olasunkanmi Adekunle Odunaiya   Department of Mechanical Engineering, Federal University Oye Ekit, Nigeria

Osato Alexendra Ighodaro   Department of Mechanical and Mechatronics, Afe Babalola University, Ado Ekiti, Nigeria

Rendani Maladzhi   Department of Mechanical Engineering, University of South Africa, Florida, South Africa

Saheed Akande   Department of Mechanical & Mechatronic Engineering, Afe Babalola University, Ado Ekiti, Nigeria

Vincent Balogun   Department of Mechanical Engineering, Afe Babalola University, Ado Ekiti, Nigeria

Wasiu Adeyemi Oke   Department of Mechanical & Mechatronic Engineering, Afe Babalola University, Ado Ekiti, Nigeria

CHAPTER 1

Introduction to the Principles of Automation and Control

Ilesanmi Afolabi Daniyan[1,*]

[1] *Department of Industrial Engineering, Tshwane University of Technology, Pretoria 0001, South Africa*

This book disseminates information about the principles and concepts of automation and control. Nowadays, more industries continue to embrace automation technologies, with the increasing use of control systems. Automation technologies and control systems find application across virtually all sectors; manufacturing economy, military, construction, and cyber security amongst others. The deployment of automation technologies boasts operational health and safety, reduction in human exposure to hazardous materials or environments, operational efficiency, time effectiveness, increase in productivity, and improvement in product quality. With automation, human error can be eliminated while repetitive or monotonous tasks are assigned to automated systems. The challenge of workers' displacement can be addressed *via* human capacity development. The upskilling of workers is a longer-term investment that can augment the expertise, skills, knowledge, and competencies of workers to enable them to collaborate effectively with machines or to advance their careers. In terms of the high initial setup cost of automating technologies, the initial cost of automating systems and processes will be offset with an economy of scale. Hence, automation also boasts a cost advantage that industries can achieve by scaling their operations, as a function of the amount of output produced. This can result in a decrease in cost per unit of output with an increase in scale. However, the major challenge of automation is the displacement of workers *via* their replacement with machines and the high initial cost that may not be effective for small to medium-scale enterprises. Chapter one provides an overview of the book while Chapters 2 to 7 are dedicated to the theoretical concepts of automation and control. Chapters 8 to 13 present ground-breaking research on automation and control and provide empirical results from the application of automation and control.

* **Corresponding author Ilesanmi Afolabi Daniyan:** Department of Industrial Engineering, Tshwane University of Technology, Pretoria 0001, South Africa; Tel: +27 (064) 5298778; E-mail: afolabiilesanmi@yahoo.com

A control system is an integral part of automation. The control system provides a means of monitoring, and tracking system's performance and execution of changes in real-time to eliminate deviations from the ideal performance. Thus control systems assist in obtaining good systems output through real-time monitoring and control. This makes many industrial processes effective and productive. This book comprises thirteen chapters that investigate the principles of automation and control.

Chapter 2 of this book presents the general introduction and definition of the basic concepts underlying automation and control. It differentiates between mechanization and automation and draws a correlation between automation and artificial intelligence. Also, the capabilities underlying AI technology are also highlighted. Furthermore, the classes of automation are explained including the procedures for automation design. The chapter ends with the merits and demerits of automation.

Chapter 3 deals with the automated processes and systems and explains the elements of system's automation. It delves into the systems operations, programming and classes of automated systems. The Programmable Logic Controller (PLC) mostly adapted for manufacturing process controls in machines, robots and assembly lines due to its merits of high reliability, ease of programming, and process fault diagnosis is also explained. The two major classes of the control system; open and closed loop control systems otherwise known as the non-feedback controls and feedback control systems respectively are discussed including their designs.

Chapter 4 discusses the levels of automation which could range from manual, semi-automatic to fully automatic depending on the level of human involvement. Furthermore, the elements of system's automation and classes of automated systems are highlighted. The identification and specifications of the elements of system's automation based on the end-user requirements are a critical aspect of the control design phase. The major elements of the system's automation include the sensor, controller, actuator, power component, motor and drives, communication protocol, human-machine interface, *etc*. Classes of automation systems could also be fixed, programmable, flexible, integrated or cognitive automation depending on the need.

Chapter 5 presents the control system and its functions, types, examples and representation of the process control systems. The chapter also discusses the types of variables: controlled, manipulated and disturbance variables. Furthermore, the types of system's processes such as batch, continuous and individual processing are explained. The chapter concludes with the Proportional-Integral-Derivative (PID) controller as a basic form of control. A PID controller is a control instrument used in industrial control applications to regulate process variables such as temperature, pressure, flow, speed, *etc*. A PID controller employs a control loop feedback mechanism to control process variables to achieve stability of the controlled variable.

Chapter 6 deals with the control devices in automation such as Programmable Logic Devices (PLD), PLC, PAC, PC *etc*. The sensors feed the main controller with the input data acquired from the environment. Following the processing of the data, a decision is made by the main controller on the control action to take and this decision is communicated to the control devices for execution.

Basically, the control devices include the input devices (for raw data input), processing devices (for processing raw data into information), output devices (to disseminate the processed data and information) and storage devices (for retention of processed data and information). The chapter concludes by differentiating between a controller and an actuator.

In Chapter 7, the emphasis is on the industrial automation tools and components. The different types of industrial automation tools such as Artificial Neural Networks (ANN), Distributed Control Systems (DCS), Human-Machine Interface (MHI), Supervisory Control and Data Acquisition Systems (SCADA), instrumentation, and robotics were highlighted. Furthermore, the application of industrial automation in robotics, packaging systems, computer numeric control systems, tool monitoring systems, advanced inspection systems as well as flexible manufacturing systems are discussed.

Chapter 8 provides practical examples of system's automation. Some specific examples presented include: the automation of irrigation system, waste segregator, gasifiers, biodiesel plantS, biogas plantS, lawn mowerS, assembly line automation as well as control and automation of railcar suspension system. The details of the design and components required for the automation of these systems are highlighted.

Chapter 9 presents a practical example of water distribution management in real time using a cloud based approach. The chapter presents the computer aided design of the proposed system as well as the materials and method necessary for achieving automation and control of this system. The performance evaluation of the developed system is discussed and the results obtained are presented.

Chapter 10 presents the automation of a waste segregator. The chapter discusses the material and method necessary for the development and implementation of an automated waste segregator including the assembly and software phases. The performance evaluation of the developed system is discussed and the results obtained are presented.

Chapter 11 presents the development of an automated fire detection and extinguishing robot. The chapter discusses the material and method necessary for the development and implementation of an automated fire detection and extinguishing robot. The highlights of the design constraints and specifications, mechanical design, motor size specification, circuitry, *etc*. were presented. The performance evaluation of the developed system is discussed and the results obtained are presented.

Chapter 12 presents the performance evaluation of a solar-powered and hand gesture-controlled lawn robot. The chapter discusses the system's architecture and design, loading requirements, as well as performance evaluation of the robot. The highlights of the design constraints and specifications, robot's components and specification, as well as circuitry *etc*. are presented. The performance evaluation of the developed system is discussed and the results obtained are presented.

Chapter 13 presents the experimental design and modelling of an automated 4-cylinder engine injector. The chapter highlights the experimental setup and modelled the working principles of a 4-cylinder injector engine. The performance evaluation of the developed system is discussed and the results obtained are presented.

CHAPTER 2

Concepts of Automation and Control

Ilesanmi Afolabi Daniyan[1,*], **Lanre Daniyan**[2], **Adefemi Adeodu**[3] and **Khumbulani Mpofu**[1]

[1] *Department of Industrial Engineering, Tshwane University of Technology, Pretoria 0001, South Africa*

[2] *Department of Instrumentation, Centre for Basic Space Science, University of Nigeria, Nsukka, Nigeria*

[3] *Department of Mechanical Engineering, University of South Africa, Florida, South Africa*

Abstract: The discussion in this chapter revolves around the general introduction and the basic definition of the concepts of automation and control. Automation and control are closely interrelated fields with the advent of Industry 4.0. Automation deals with the integration of technologies that can enable systems to carry out tasks without human intervention or with minimal intervention. On the other hand, control is a process of monitoring and manipulating the variables of a system in order to achieve the desired outputs. Hence, an automated system comprises the control system, information, communication and technology system, actuator, and effective feedback mechanism. The emergence of Industry 4.0 technologies focuses on improvement in efficiency, profitability, systems' flexibility, manufacturing processes, product quality, cost, and time effectiveness with a significant reduction in manufacturing process errors. These improvements can be aided by putting in place a system with effective automation and control. This chapter further explores the differences between mechanization and automation and draws a correlation between automation and artificial intelligence. Also, the capabilities underlying the Artificial Intelligence (AI) technology are highlighted. Furthermore, the classes of automation are explained including the procedures for automation design. In addition, the merits and demerits of automation are highlighted and the chapter ends with the automation of production lines and different work layout configurations. The concept of automation is central to industrial society and is prevalent in the engineering industries (manufacturing, process industries, *etc.*).

Keywords: Artificial Intelligence, Automation, Control, Industry 4.0.

GENERAL INTRODUCTION AND DEFINITION OF BASIC CONCEPTS

The word "automation" is derived from the Greek words "Auto" (self) and "Matos" (moving). Therefore, automation means the development of a mecha-

* **Corresponding author Ilesanmi Afolabi Daniyan:** Department of Industrial Engineering, Tshwane University of Technology, Pretoria 0001, South Africa; Tel: +27 (064) 5298778; E-mail: afolabiilesanmi@yahoo.com

nism for systems to operate by themselves. Hence, systems are automated to move, adjust and implement instructions by themselves. It is a technological method by which a process or system is controlled with the use of electronic, mechanical and computer-based instructions without human intervention [1]. It could also be defined as a set of technologies integrated to enhance machine independence during operation without significant human intervention or the application of machines to tasks once performed by human beings or, to tasks that would otherwise be impossible to perform by humans [1].

The control of a process or system by automatic means rather than manual is often called automation. It comprises a set of technologies by which simple or complex processes or systems can be operated independently or with little human intervention. The set of technologies are integrated into a self-governing system for the execution of a particular task. Systems are automated to minimise their interactions or dependencies on human personnel [2]. A control system is a set of technologies used to adjust the process parameters to achieve the desired output. Therefore, through effective control, the desired output of a system can be achieved by adjusting or regulating the input variables [3]. Hence, automation and control refer to the collection of personnel, hardware and software employed to ensure effective monitoring, precision and accuracy, safety, security, efficiency, productivity and reliability of the manufacturing or industrial process [4]. The automation systems encompass the control system, information, communication and technology system, actuator and effective feedback mechanism.

The concept of automation is central to industrial society and is prevalent in the engineering industries (manufacturing, process industries, *etc.*). To reduce the rising wages and its associated production cost, automatic machines are employed to increase the production of a plant per worker. Manufacturing industries are profit-oriented and are concerned with the precision and productivity per worker of their plants. Automatic systems offer the solution of high productivity without sacrificing precision and accuracy [4]. Hence, the reduction of human interference in the operation of machines and direct replacement by technologically driven systems, such as computers, robots *etc.* is referred to as automation.

The performance of automated systems in terms of accuracy, precision, speed of operation, and productivity, is usually superior when compared to manual systems. Automation covers a broad range of applications ranging from simple systems such as household devices to complex industrial systems. For large and complex systems, there are thousands of input variables and output signals which are measured and controlled autonomously to enhance the system's independence. The control may be in the form of a simple ON/OFF control to complex or multivariable high-level algorithms. Industrial automation utilizes control systems as well as information technologies to handle different processes and machinery in the industries. Automation has helped in improving production quality and quantity, thus, making production lines much more flexible. The technology can be deployed for material handling operations, assembly, production, machining, transportation, inspection, quality assurance and packaging amongst others [5]. A robust automation system often entails control technologies, artificial intelligence, enabling communication protocols, and hardware sections. This will assist manufacturing industries to gain a competitive edge, ensure production system's reliability and promote high production efficiency. It will also assist in meeting the challenges of the increasing dynamics and complexities of manufacturing and product development. Automation and control are terms often used together or interchangeably. Control involves the operation or adjustment of devices or components in order to ensure that the system does not deviate from the set or desired points. These adjustments are done with the aid of control systems, devices or actuators. Common examples include the turning ON and OFF of light and the use of the press button of wireless remote controls. Many other systems in and outdoors also possess devices for effective control such as indoor and outdoor lighting systems, air condition systems, television, cooking and refrigeration system. The elements of control and actuators are a subset of automation. On the other hand, the process of automation integrates many control devices with effective interaction to carry out many tasks independently with a centralized intelligent control system that responds to input signals in order to control each of the control devices. With the integration of many control devices controlled centrally with an intelligent control system, the overall system can be prog-rammed to run with the least human intervention. The branch of engineering which uses programmable machines to automate activities is referred to as robotics. The programmable machines are called robots. Depending on the system's requirements and the level of automation, robots can operate autonomously or semi-autonomously. They are designed to interact and interface with the physical world with the aid of sensors, cameras and actuators with good learning and perception abilities of the environment. The ones that are reprogrammable are flexible enough to permit dynamic changes and robots can be collaborative in nature permitting activities to be carried out with humans at the same time. Automation can be achieved with the combination of mechanical, electro-mechanical, electrical, and electronic devices, computer and computer programs, as well as pneumatic and hydraulic systems.

MECHANIZATION AND AUTOMATION

In the scope of industrialization, automation is beyond mechanization. While mechanization is a vital component of industrialization replacing human drudgery with mechanical devices, leading to high productivity, better working conditions of personnel's, and in general, more profit for entrepreneurs with significant human involvement, automation deals with the deployment of technology to make systems work independently with minimal human interference. Nowadays, many mechanized processes are being upgraded to include the decision-making attributes of human beings, which is regarded as automation. While mechanization has contributed a lot to reducing physical labor, automation helps in reducing the mental labor in production. Encyclopedia Britannica defines the term "mechanization" as the use of machines to replace human labour, while automation generally refers to the integration of the mechanical, electro-mechanical, electrical, electronic, embedded and computer systems as well as the control systems and information technology into a self-governing system for efficient actuation, monitoring and control.

AUTOMATION AND ARTIFICIAL INTELLIGENCE

The world of automation has continued to advance at a fast pace due to techno-logical advancement. Automation involves the integration of hardware, software and information system to enable a machine to function independently while Artificial Intelligence (AI) otherwise called machine intelligence is a science that deals with the development of systems or machines with cognitive ability and intelligence which range from simple to human-like intelligence. The intelligence of a machine enables it to rationalise and take decisions and actions that will enhance production goals and objectives. Machine intelligence is developed to replicate human intelligence much the same way the human brain, thinks and functions. It, however, relies on mathematical algorithms, data sets, certain features or hidden patterns to build a predictive model after an iterative training of the data set or features extraction.

The following are the capabilities underlying AI technology:

1. Problem-solving: Smart machines and systems are incorporated with algorithms that enhance quick problem-solving *via* the imitation of human reasoning.

2. Machine learning: This involves the use of algorithms and statistical models to acquire, study and retain information. The process of acquiring large datasets and extracting information for decision-making is called data mining. Hence, machine learning is a form of data mining technique to develop model equations, and decisions are often made based on what the machines had learned. A machine can learn through different methods ranging from supervised to unsupervised learning. Depending on the nature of the problem to be solved, there are different algorithms such as regression and clustering algorithms which can be employed for prediction, classification, pattern recognition and regression problems. The

data employed for training is referred to as the training set comprising the input variables (independent variables) and the output variables (dependent variables). The process of training is the period the machine takes to study the relationship between the independent and dependent variables. The process of training is followed by the development of a model which is tested with the use of a new dataset called the test data set. The model can subsequently be deployed for use if it shows evidence of good predictive and correlative abilities. Deep learning is a subset of machine learning which involves the structuring of algorithms in different layers to create an artificial neural network, which can be used for learning, pattern recognition and decision-making. The neural network is designed to imitate the biological neurons and can study information and recognise patterns. Examples of machine learning include; supervised learning, pattern recognition, *etc*.

3. Language processing: Smart systems are designed to study and understand human language in order to facilitate effective communication between the system and the operator. The system possesses translation and communication abilities with the aid of signal processing and semantic and syntactic analysis.

4. Motion and perception: This is the sensual ability of the system that enables it to move and relate to the environment. Smart systems are incorporated with sensors and camera functions to aid easy perception and interaction with the environment.

5. Creativity: This is the ability of the machine to think and act uniquely.

CLASSES OF AUTOMATION

Depending on the sections of production to be automated and their scopes, automation may be classified into three namely: Industrial, software and process automation.

Industrial Automation

Industrial automation of a process or system involves the application of process control and information systems. Industrial automation is a vast field which that involves the integration of process control, machines, electronics, software, and information systems to achieve improved product quality, increased production and production flexibility at optimum cost. This involves the integration of the machines, process, IT and communication systems on the production floors. In other words, it involves the use of physical machines and control systems to autonomously perform an industrial task. It encompasses the automation of material handling processes, manufacturing techniques as well as quality control processes [6]. The goal of industrial automation is to use Information Technology and communication system to enhance the independent operation of the technical process [7]. This includes both hard/mechanized automation (such as automation of milling processes on a manual machine using different jigs and fixtures to support the tools and work piece) and soft/programmable automation – which employs the use of programmable logic controllers (PLCs), CNC machines, robots and so on. Industrial automation is often focused on developing embedded systems to automate tedious or repetitive manufacturing tasks by providing appropriate materials handling technologies, and flexible jigs and fixtures to support the precise location of the work piece and the tools. Discrete control activities that may be carried out include sequential control, speed control, packaging and batch control. Actuators such as magnetic valves, on/off drives/ motors, and limit/proximity switches are used along with micro-controllers and modular PLCs as hardware for manufacturing automation. The structure of industrial automation includes the following;

1. Operators: they are the personnel who interact with the central control system, andoperate and manage the process.

2. Information technology and communication system: for effective communi-cation and control of the system, the information technology and communication system provides capabilities for the entire system to enhance its independence; for instance the use of artificial intelligence for process improvement, real-time processing such as high-performance computing, cloud computing, artificial vision, advanced users' interface, advanced programming such as C, C++, Python, Java, Perl, NET.

3. Technical system: This is the system where the transformation of the material and energy takes place.

An industrial automation system can be regarded as a real-time system. A real-time system is a system, which has the capability of processing incoming data continuously with defined time constraints so that the output results are available within a given period of time. The task of a real-time system is usually defined in three states namely; running (which implies the execution of the task on the central processing unit), ready (which means that the task is ready for execution) and blocked (which denotes that an event is awaited). It has the hardware/software system as well as the data acquisition, processing, and delivery system for onward transmission of information to the process within a given time interval. The following are the characteristics of a real-time system;

1. Time bound: It can react within the fixed time constraint.

2. Versatility: It can simultaneously react to various events and activities within the system.

3. Reliability: Its reaction is usually accurate and precise.

4. Predictability: All reactions within the system are predictable and deterministic.

Software Automation

This involves the use of software to perform tasks and activities using computer programs. The types of software automation include;

1. Business process automation: This involves the structuring of business processes and their integration into the automation software.

2. Robotic process automation: The use of robots that are programmed to use computer programs for task execution.

3. Intelligent process automation: This involves the use of artificial intelligence to perform tasks using computer programs.

Process Automation

A process is a combination of activities that require the flow and transformation of materials, energy or information from one state to another state. Process automation involves techniques for continuous monitoring and controlling of measurements during a process [8]. Process parameters are measured using various field-sensing devices such as process temperature, pressure, flow, level, vibration, switches, and control valves. Control actions in process automation have a higher delay time factor when compared with manufacturing automation [9].

If the processes or tasks to be performed are in a virtual environment, software automation will be a viable option but if such activities involve physical machines and other elements, industrial automation will be more suitable.

PROCEDURES FOR AUTOMATION DESIGN

The following are standard procedures for the design of automation processes:

1. Establishment of the system requirements.

2. Analysis of the system capabilities and development of a formal specification of the system's performance.

3. Development of a dynamic model of the system, often with the use of system identification techniques.

4. Development of the control system.

5. Simulation of the system to check the effect of the controller on it.

6. Acquisition of the necessary hardware for sensing, actuating, control implementation, information communication and artificial intelligence functions.

The processor suitable for industrial and process automation will have a programmable digital processor which is compliant with the requirements of real-time operation requirements (*i.e* having the capacity for the acquisition, processing and delivery of output data in a timely manner). In addition, it should have the capabilities for the connection of input/output of process signals either directly or *via* a communication system and should be capable of processing numbers, characters and bits.

MERITS AND DEMERITS OF AUTOMATION

The process of automating systems and subsystems has a wide range of advantages and disadvantages. To fully harness the advantages of the system's automation, it is pertinent to decide on the level of automation which best suits the system based on the expertise of the personnel, as well as the cost-benefit analysis of the entire system.

Merits of Automation

The following are some of the merits of automation [10-12].

1. It increases the quality, conformity and uniformity of products through effective monitoring and control.

2. It increases the reliability and speed of the production process thereby reducing production cycle time.

3. It permits high reproducibility and is suitable for monotonous and repetitive tasks.

4. It reduces the cost of production in the long run.

5. It increases production monitoring and control.

6. It enhances the quality of life by providing efficient operation, a safe working environment and a reduction in man-machine interaction.

7. Increased output, productivity and efficiency.

8. Reduce direct human labour as well as its associated costs.

9. Creation of higher value jobs to support and maintain industrial control system environments.

10. Permits proper data gathering, computation and management which is key to process improvement.

11. It permits quick response to signals and is suitable for handling complex operations.

12. It extends the equipment service life, and saves consumables and raw materials.

Demerits of Automation

Some of the disadvantages of automation are as follows:

1. It makes the production system inflexible to changes. As a process becomes increasingly automated, there is less quality improvement to be gained.

2. It results in the replacement of specialized personnel with machines thereby causing unemployment.

3. High initial cost. The cost of procuring automated systems as well as the cost incurred for the research and development of automating a process may exceed the cost saved by the automation itself. Switching from manual production lines to automatic as well as training of personnel to handle automatic production lines is usually expensive. Hence, there is a need for a cost-benefit analysis of the automation technology to justify the initial cost of investment.

4. There are some industrial tasks that require human reasoning and judgment, pattern recognition, strategic planning, and language comprehension assessment of sensory data such as scents, odour, and sound that require human expertise. In addition, there may be a need to improvise and develop some flexible procedures manually. All these may not be possible in a fully autonomous system.

5. Sometimes, the use of inappropriate technology or inappropriate design for automation and control may increase the chances of error generation and subsequently system's failure.

6. At the intermediate level of automation, where there is a need for human-system's collaboration, a lack of proper understanding of the automation tech-

nology by the personnel may increase the chances of error generation and subsequently system failure. This challenge however can be solved *via* the development of human capacity and skills training on the utilization and effective collaborations with the control systems.

From the merits and demerits of automation highlighted above, the robustness and flexibility in system or process automation can only be achieved *via* effective human-machine collaboration with proper task allocation.

AUTOMATED PRODUCTION/ASSEMBLY/ FLOW LINES

Automated flow lines comprise several workstations or machines integrated such that there is a transfer of parts between the stations. A workstation is a group of machines or operators that perform one or more operations on the job. One of the design requirements of automated production lines is the development of minimal capacity in terms of the number of machines or operators that are required to achieve a certain target throughput level. Other design requirements include the degree of parallel workstations, the location and size of buffers in between the workstations, the choice of the material handling system, tasks allocation to the work stations, as well as the assignment of operators to the workstations. Production lines may be divided into three groups: synchronous, asynchronous and continuous lines. In synchronous lines, the job movement along the work-station is simultaneous and coordinated, thus, the number of jobs in the system remains constant, and there is no need to locate buffers in between stations.

In asynchronous lines, the movement of jobs is not simultaneous and coordinated. The operator or machine starts to process the next job as soon as the job becomes available. On completion of the job, it is moved immediately to the next work station, as long as there is space for it. In continuous lines, there is an uninter-rupted flow of jobs and the aim is to produce a flow production to manufacture, produce, or process materials without any interruption.

The aim of the automated flow lines is to integrate operations to achieve a reduction in work-in-process, and labour costs as well as an increase in quality and production rates with minimum movements between operations.

TYPES OF WORKSTATIONS LAYOUT/CONFIGURATIONS

There are three major types of automated workstation configurations namely: the in-line or straight configuration, segmented or cellular configuration and the fixed configuration.

In-Line or Straight Configuration

For the inline or straight configuration, the machines or workstations are arranged sequentially in a straight line. This type of configuration is also referred to as the product layout and is suitable for repetitive assembly or process or in continuous flow industries. The straight line configuration is more effective for the production of a repetitive, dedicated, high volume and standardized product.

This layout allows the entire manufacturing process to be laid out in a straight line, which at times may be totally dedicated to the production of only one product or product version. The two types of lines commonly used in the straight configuration are paced and un-paced. Paced lines can make use of a conveyor for moving parts or products along a straight and continuous line so that operators can perform operations as the part or product is conveyed.

For the unpaced line, workers queue up between workstations to permit a variable work pace. The assembly-line balancing technique can be used to allocate tasks to the different workstations at equal time requirements. With the line balancing technique, individual tasks performed can be grouped into the workstations to enhance a balance of work among the workstations.

The inline or straight configuration has the following merits:

1. Output and time effectiveness: It is suitable for generating high volume of products in a short time.

2. Cost-effectiveness. The unit cost is reduced due to high volume production.

3. Specialization of labour: Due to the fact that the task is routine and repetitive, there is job specialization which often results in reduced training time and cost.

4. Machine utilization. There is a high degree of labor and equipment utilization.

The disadvantages of the inline or straight configuration include:

1. Reduction in employee motivation: The division of labour which characterize this configuration can result in dull, repetitive jobs that could be stressful.

2. Inflexibility: The inline or straight configuration is relatively inflexible and cannot easily respond to required system changes—especially changes in product or process design.

3. Increase in downtime: The manufacturing system is at risk of equipment breakdown, employee absenteeism, and downtime due to preventive maintenance.

Segmented or Cellular Configuration

In the segmented or cellular arrangement, the machines or workstations are arranged according to the process requirements for a set of part families (similar items) that require similar processing. The groups of machines are called cells. The arrangement of the machine can take an L or U or rectangular shape.

Similar processes can be grouped into cells using a technique known as group technology. The group technology entails the identification of parts with similar design characteristics such as shape, size and function and similar process characteristics such as the type of processing required, available processing machines and the processing sequence.

An automated form of cellular manufacturing is called the Flexible Manufacturing System (FMS). For the FMS, the transfer of parts to the various processes is achieved with the aid of computer controls.

The cellular configuration boasts of the following merits [13-15].

1. Product quality and easy rework: The cellular configuration contributes to the development of products with good quality. Once a defect is identified in the flow line, the product can easily be returned to the workstation where such product is manufactured in a time-effective manner. This is because the distance and time taken to return a defective product to the workstation where it is manufactured are shorter compared to the conventional flow line where a defective product can be sent to a separate workstation for rework.

2. Cost. Cellular configuration provides for faster processing time, with less work-in-process inventory, and reduced setup time. All these contribute to significant cost savings.

3. Volume flexibility: Cellular configuration allows for batch production. The volume of the workstation is flexible and can be reconfigured or adjusted to meet the current demand in line with the JIT manufacturing principles. Furthermore, the production rate as well as the number of personnel on each of the workstations can also be adjusted to suit the current demand compared to other flow line configurations.

4. Number of workstations: It is easier to group activities and personnel into various workstations in the cellular configuration compared to other configurations. This makes it easier to achieve a reduction in the number of workstations without compromising on the product quality as well as manufacturing time and efficiency.

5. Supervision and operators' flexibility: Supervisory duties are better carried out in cellular configuration since the distance from one workstation to another is relatively short compared to other configurations. Thus, the operations of the personnel become more flexible sometimes, an operator can multi-task among the various workstations.

6. Material handling: In the cellular configuration, the need for material handling equipment such as conveyors and operators are minimal since products are moved directly from machine to machine or workstation to workstation.

7. Compactness, visibility and teamwork: Cellular configuration is more compact than other configurations with limited space wasted in line with the 5S principles (Sort, Set in order, Shine, Standardize and Sustain). The arrangement of the cellular configuration whereby the machines and workstations are close to each other visibility and communication of the employees compared to a straight line where operators are spread out along a long line and may be separated by different machines or workstations. For instance, the space at the centre of the U-flow line is a shared area where operators can display, communicate, assist one another, and learn from others.

8. Waste reduction: Waste generation can be a result of any of the following: defects, waiting time, extra motion, excess inventory, overproduction, extra processing, unnecessary transportation or as a result of potentials that are not maximally harnessed for manufacturing. In line with the 5S principles, geared towards waste reduction, tools, machines or workstations are better sorted, and set in order in the U-shaped configuration compared to other configurations.

9. Motivation: The motivation of workers also increases with a reduction in boredom, and monotonous, or repetitive tasks.

Fixed Layout Configuration

A fixed-position configuration is suitable for products that are too heavy to move from one workstation to another. For instance, battleships are not produced on an assembly line rather they are produced in fixed production lines. Other construction or manufacturing activities relating to aircraft, drilling *etc*. are carried out on fixed production lines.

One of the major disadvantages of the fixed layout configuration is space.

A properly designed assembly or flow line is important for mass production. Many manufacturing industries are implementing the Just-in-Time (JIT) manufacturing model whereby products are developed to meet demand, not developed in advance or surplus. The objective of the JIT is to increase manufacturing lead time and efficiency at an optimum cost. With the implementation of the JIT manufacturing principles, many industries adopt the segmented, cellular (U-shaped) assembly or flow time rather than the conventional straight lines [16].

CONCLUSION

The advent of the Fourth Industrial Revolution driven by cyber physical systems makes automation and control inescapable in the industry and in modern society. Almost every complex engineering system; automobile, rail, aircraft, home or building appliances, oil refinery or other process plants, robots *etc.* requires some level of automation to function independently. The safety, precision and efficiency of these systems are also a function of the control system for sensing, feedback, and actuation. Hence, automation and control promote the digitalisation and the building of a virtual environment heralded by the Fourth Industrial Revolution. Industries will need increased technological advancement through increased automation and control to meet the bottom line goals of sustainability, profitability, competitiveness and market share.

REFERENCES

[1]   Introduction to industrial automation and control. Lesson 1. Version 2 EE IIT, Kharagpur Available at: https://nptel.ac.in/content/storage2/courses/108105063/pdf/L-01(SM)(IA&C)%20((EE)NPTEL) .pdf Accessed 17th February, 2022.

[2]   I. Oditis, and J. Bicevskis, "The concept of automated process control", In: *Scientific Papers* vol. 756. University of Latvia, 2010, pp. 193-203.

[3]   S.H. Saeed, *Automatic control systems.* Kataria & Sons Publisher: New Delhi, 2008.

[4]   H. Eren, and C-C. Fung, "Automation and control equipment and implementation", In: *Wiley Encyclopedia of Electrical and Electronics Engineering.,* J. Webster, Ed., John Wiley & Sons, Inc., 1999, pp. 146-164.

[5]   B. Connell, *Process Instrumentation Applications Manual.* McGraw-Hill: New York, 1996.

[6]   D.M. Considine, and G.D. Considine, *Standard Handbook of Industrial Automation.* Chapman & Hall: New York, 1984.

[7]   B.J. Kuo, *Automatic control systems.* 6th ed.. Prentice-Hall: Englewood Cliffs, NJ, 1991.

[8]   W.S. Levine, *The Control Handbook* Boca Raton, FL: CRC Press, 1996.I. Nagy, *Introduction to Chemical Process Instrumentation* Elsevier: New York, 1992.

[9]   K. Ogata, *Modern Control Engineering.* 3rd ed.. Prentice-Hall: Upper Saddle River, NJ, 1997.

[10]   M.J. Pitt, and P.E. Preece, *Instrumentation and Automation in Process Control.* Ellis Harwood: Chichester, England, 1991.

[11]   J. Webb, and K. Greshock, *Industrial Control Electronics.* 2nd ed.. Macmillan: New York, 1993.

[12]   L.C. Westphal, *Sourcebook of Control Systems Engineering.* Chapman & Hall: Cambridge, 1995. [http://dx.doi.org/10.1007/978-1-4615-1805-1] 12B. Maskell, *Performance Measurement for World Class Manufacturing.* Productivity Press, Inc.: Cambridge, MA, 1991.

[13]   G.J. Miltenburg, and J. Wijngaard, "The U-line Line Balancing Problem", *Manage. Sci.,* vol. 40, no.

10, pp. 1378-1388, 1994.
[http://dx.doi.org/10.1287/mnsc.40.10.1378]

[14]   Y. Monden, *Toyota Production System.* 2nd ed. Industrial Engineering Press, Institute of Industrial
       Engineers: Norcross, GA, 1993.
       [http://dx.doi.org/10.1007/978-1-4615-9714-8]

[15]   M.J. Schneiderjans, *Topics in Just-In-Time Management.* Allyn and Bacon: Needham Heights, MA,
       1993.

[16]   R.J. Schonberger, and E.M. Knod, *Operations Management: Continuous Improvement.* Irwin: Burr
       Ridge, Ill., 1994.

CHAPTER 3

Automated Processes and Systems

Saheed Akande[1,*], **Wasiu Adeyemi Oke**[1] and **Fawaz Aremu Babajide**[1]

[1] *Department of Mechanical & Mechatronic Engineering, Afe Babalola University, Ado Ekiti, Nigeria*

Abstract: In this chapter, the concepts of automated processes and systems including the elements of system automation are explained. In process automation, digital technologies are often used to automate complex manufacturing or business processes. This includes the use of the system to perform tasks and the integration of software, information and communication technology, data acquisition, and storage sub-systems. A system may go through several processes that are time sensitive and repetitive before obtaining the final output. Process automation prevents variation and bottlenecks associated with these processes such as errors, and data loss while improving speed and communication among other sub-systems. System automation is a subset of Mechatronics engineering that involves the integration of a sensing system, control mechanisms, and drive system (actuators). The three basic elements of an automation system that must be synergized include: the power system, the program of instructions or codes to direct the process, and the control mechanism to actuate the instructions. This chapter also delves into the systems operations, programming and classes of automated systems. The two major classes of the control system; open and closed loop control systems otherwise known as the non-feedback and feedback control systems respectively are discussed including their designs. The Proportional Integral Derivative (PID) controllers will continue to be important in several industrial applications because they utilize a control loop feedback mechanism to control process variables and are highly stable and accurate in achieving control tasks.

Keywords: Automated processes, control system, PID, system automation.

INTRODUCTION

Automated solutions are increasingly used in the manufacturing industries because they can be deployed to replace slow manual processes in order to achieve production systems with higher efficiency. The streamlining of processes, workflows, and management of complex manufacturing technologies can also be achieved via automation. This chapter discusses the control system automation, classes of automation, basic controls, and the control system in general.

---
[*] **Corresponding author Saheed Akande:** Department of Mechanical & Mechatronic Engineering, Afe Babalola University, Ado Ekiti, Nigeria; Tel: +2348061231291; E-mail: sakande1@abuad.edu.ng

Ilesanmi Afolabi Daniyan (Ed.)

AUTOMATION PROCESSES

The word 'Automation' is derived from the Greek word "Auto" which means self and "Matos" which means motion. Automation is a technology that involves the application of mechanical, electrical, electronics, computer hardware, and software systems to operate, control and guide production processes and systems without significant human intervention and achieves performance significantly better than manual operation [1]. It is a subset of Mechatronics Engineering. Automations are widely used in almost all phases of production, industrial, business as well as the domestic world [2]. The application of automated systems and processes is unlimited ranging from domestic vacuum cleaner (domestic robot) to sophisticated machine assembling plant, automatic inspection system, and an industrial robot that operates in environments not suitable for human beings like high-temperature furnaces, and high-frequency electromagnetic waves (like gamma rays and X-rays) environments. Also, in the banking and business world, the use of automation is inevitable. Automated Teller Machine (ATM), and automatic money counter with fake note detectors among others are typical applications of automation systems.

ELEMENTS OF SYSTEM AUTOMATION

System automation is a subset of Mechatronics engineering and it involves the integration of sensing systems, control systems, and drive systems (actuators). The three basic elements of an automation system [3] are:

1. The power to accomplish the process and operate the system.

2. A program of instructions (code) to direct the process,

3. A control mechanism to actuate the instructions.

The relationship between these elements is illustrated in Fig. (**1**).

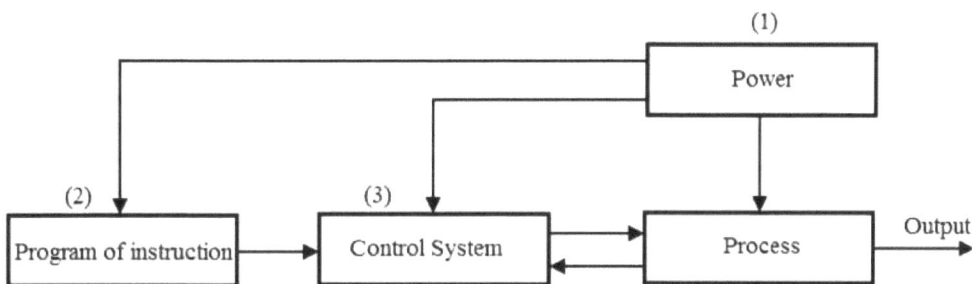

Fig. (1). Relationship among the elements of system automation.

THE POWER TO ACCOMPLISH THE PROCESS AND OPERATE THE SYSTEM

The output of the automation system is motion or action. Thus, energy is required. Fundamentally, the energy expended per unit of time is power. Therefore, power is also required in an automation system. Just like different types of energies are available, power is also of different forms. Electrical power is the most commonly used form of power to operate automation systems due to its availability, cost, and ease of conversion to other forms like mechanical, heat, and light energy among others. The two main functions of power as an element of automation are: the accomplishment of the automation process and the operation of the automation system [4]. Electrical powers at low levels are required for the accomplishment of automation processes like signal transmission, information processing, data storage, and communication [5].

The output of the automation system (motion) performs two major functions;

i. Processing: These include shaping, molding, loading, and unloading activities. Energy is applied to accomplish processing operations on some entity and thus power is required to convert the entity from one state into more required and valuable forms.
ii. Transfer and position: This involves the moving of the product from one location to another during the series of processing steps, the positioning and placements of the products for processing, *etc.*

PROGRAM OF INSTRUCTIONS

This is a set of commands that specify and define the sequence of steps or actions to be performed in an automation system. It is the foundation and the heart of an automation system [6]. A common example of a program of instruction as an element of automation is the use of Computer Numerical Control (CNC) for automating machine tools movements that control auxiliary functions including spindle through the use of software (code) embedded in the microcomputer attached to the tool [7]. The program written in an international standard language called G-code contains the instructions and parameters the machine tool will follow [8]. Milling, laser cutting, and lathe among others are common machine tools that can be automated with CNC. Some non-machine tools' operations like welding, assembly, and disassembly can also be automated on CNC. Each operation has a custom computer program, usually stored and executed by the Machine Control Unit (MCU). Other examples of the program of instruction include the temperature setting of a furnace, the ON and OFF of an electric motor, the use of an automated Universal Testing Machine (UTM), *etc.*

A CONTROL MECHANISM TO ACTUATE THE INSTRUCTION

The control element of the automated system executes the program of instructions to ensure that the process accomplishes its defined function. There are basically two types of control systems in automation; the open loop and the closed loop systems [8].

THE TRANSFER FUNCTION

In order to understand the aforementioned two types of control systems, it is necessary to briefly introduce a function that provides a relationship between the input and the output of the system. This function is called the transfer function. The transfer function is defined as the ratio of the Laplace transform of output to the ratio of the Laplace transform of the input with the assumption that the initial condition is zero. Hence, the transfer function of a control system is defined as:

$$G(s) = \frac{\text{LaplaceTransformofOutput}}{\text{LaplaceTransformofinput}} = \frac{Y(s)}{X(S)} \qquad (1)$$

Open-loop System

This is also called the non-feedback system. It is a type of continuous control system in which the output has no influence or effect on the control action of the input signal. In such a system, the output is neither measured nor "fed back" for comparison with the input. Therefore, an open-loop system follows its input command or set point independent of the final result. Also, it has no knowledge of the output condition so cannot self-correct any errors it could make when the pre-set value drifts, even if this results in large deviations from the pre-set value. Fig. (**2**) shows a typical open-loop system circuit.

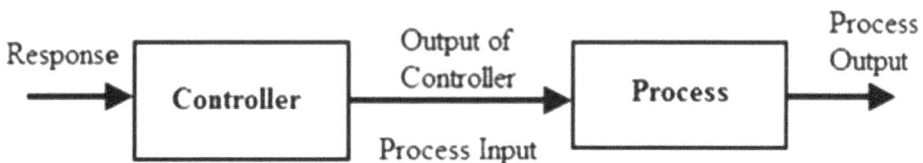

Fig. (2). Open loop system circuit.

Examples of open loop systems include clothes dryers, electric bulbs, TV remote control, washing machine, servo motor or stepper motor, door lock systems, and

Inkjet printers among others. In a cloth dryer, for example, the control action is carried out by the operator by fixing the timer for say 10 minutes. Thus, after 10 minutes, the system will stop even if the cloth is wet. This is because there is no feedback and the controller of such a system is the timer. Fig. (3) displays the block diagram of a cloth dryer.

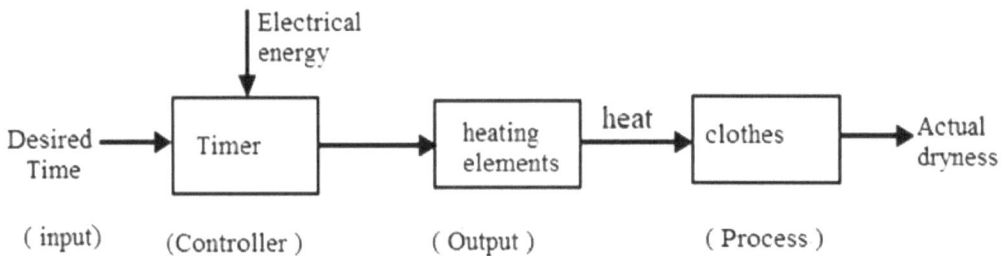

Fig. (3). The block diagram of clothe dryer.

Another disadvantage of open-loop systems is that they are poorly equipped to handle disturbances or changes in the conditions which may reduce their ability to complete the desired task. For example, if the dryer door is opened and heat is lost, the timing controller continues regardless of the set 10 minutes, though the cloth remains wet. This is because there is no information fed back to maintain a constant temperature.

To calculate the transfer function of an open loop control system, the relationships or equations from each block diagram are considered. Consider an open loop system in Fig. (4).

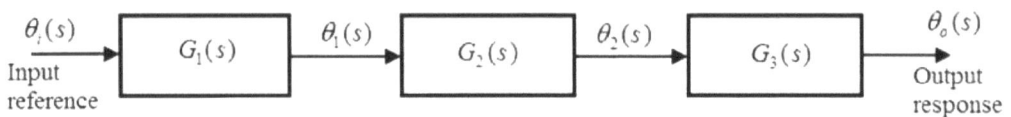

Fig. (4). Transfer function of an open loop control system.

The transfer function G(s) of an open loop system can be computed by combining the transfer function for each of the blocks using Eq. (1) as follows:

$$G_{1(s)} = \frac{\theta_1(s)}{\theta_{i(S)}},$$

(2)

$$G_{2(s)} = \frac{\theta_2(s)}{\theta_1(s)}, \tag{3}$$

$$G_{3(s)} = \frac{\theta_{o(s)}}{\theta_2(s)}. \tag{4}$$

Then, the overall transfer function G(s) is the product of all the blocks transfer functions, that is,

$$G(s) = G_1(s) \times G_2(s) \times G_3(s) = \frac{\theta_1(s)}{\theta_{i(S)}} \times \frac{\theta_2(s)}{\theta_1(s_-} \times \frac{\theta_{o(s)}}{\theta_2(s)} = \frac{\theta_o(s)}{\theta_1(s)}$$

And

$$G(s) = \frac{\theta_o(s)}{\theta_1(s)}. \tag{5}$$

A Closed-loop Control System

This is also known as a feedback control system. In this case, the output variable is compared with an input parameter, and any difference between the two is used to drive the output into agreement with the input. Closed-loop systems are simply complete automation control systems because they are designed in such a way that the achieved output is automatically compared with the reference input to determine the required output. Fig. (**5**) illustrates the block diagram of a closed-loop automation system.

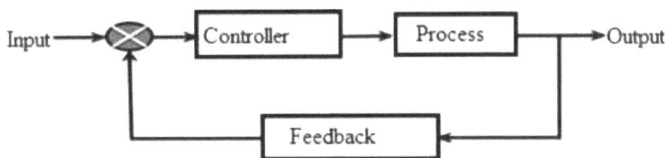

Fig. (5). The block diagram of a closed loop automation system.

In a **closed-loop system**, the feedback loop (as a part of the output) is a key parameter directed to the input signal so that both the input and the output can be compared and the desired output can be achieved if the present output shows variation in the desired output. Thus, it can be said that the output performs the controlling action in a closed-loop system. There are two basic forms of feedback in any automation circuit; Positive and Negative feedback.

i. Positive Feedback

This is the type of feedback in a control system in which the input signal and the feedback signal are in phase with each other. In these systems, the reference input is added to the feedback signal thereby increasing the gain of the overall system. Fig. (6) displays the block diagram of the positive feedback control system.

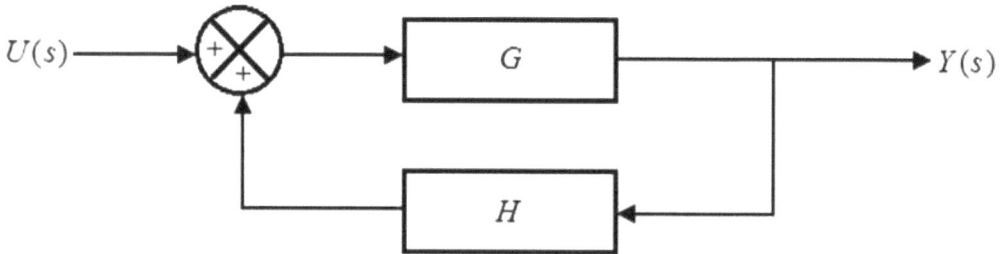

Fig. (6). The positive feedback control system.

The transfer function $G_c(s)$ of the positive feedback control system is given as:

$$G_c(s) = \frac{Y(s)}{U(s)} = \frac{G}{1-GH} \tag{6}$$

Where G is called the open loop gain and H is called the gain of the feedback path.

ii. Negative Feedback

In this case, the input signal and the feedback signal show an out-of-phase relationship with respect to each other. Thus the applied input signal and the feedback signal are subtracted to get the error signal. This leads to a reduction in the overall gain of the system. The block diagram of the negative feedback control system is shown in Fig. (7).

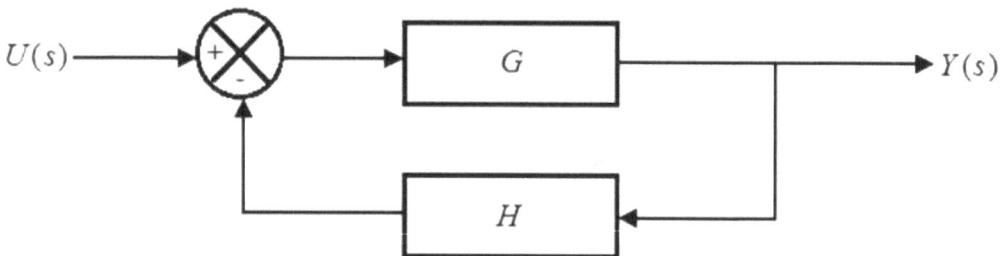

Fig. (7). The negative feedback control system.

Transfer function $G_c(s)$ of the negative feedback control system is defined as:

$$G_c(s) = \frac{Y(s)}{U(s)} = \frac{G}{1+GH} \qquad (7)$$

Where:

G and H are as defined in Eq. (6).

Examples of closed loop control systems include an air conditioning system, voltage stabilizers, thermostat heaters, electric Iron, water level controller, automatic street light, and smoke detection systemd among others.

Consider for example, an automatic electric iron closed-loop system. It consists of a thermostat that acts as a controller of the system, a resistive heating element that generates heat while the sole-plate of the iron acts as a process of the overall system. The control of automatic electric iron is such that when the temperature of the sole plate attains a predefined value, the heating ceases automatically. However, when the temperature falls below a specified threshold value, the heating commences. Fig. (**8**) represents the block diagram with major compo-nents.

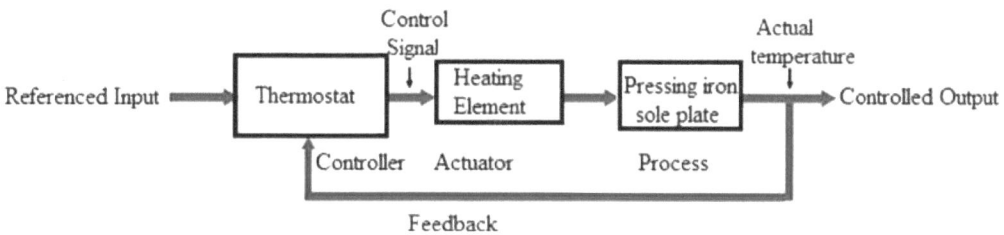

Fig. (8). Block diagram of a closed system of an automatic electric iron.

CLASSES OF AUTOMATED SYSTEMS

There are different classes of automation systems and the choice of a particular method depends on the nature of the output. The major common ones are:

Fixed or Hard Automation

This is the type of automation that carries out a single set of tasks without deviation. It involves the integration and coordination of many operations into one piece of equipment that makes the system complex [3]. The system is used for discrete mass production and continuous flow systems. Typical examples of fixed automation operations include machining transfer lines in the automotive industry, conveyors, automatic assembly machines; and certain chemical processes like paint production, distillation process, and solvent extractions, among others. Fixed automation requires high initial investment and a high production rate though the process is relatively not flexible to accommodate changes in product design. The economic justification for fixed automation is found in the products with very high demand rates and volumes.

Programmable Automation

In this case, the system is designed in such a way that it can be altered to accommodate the change in the sequence of operations for different product configurations using electronic controls. The operation sequence is controlled by a program, which is a set of instructions coded so that the system can read, understand, interpret, and carry out the instruction. When a new product is required, new programs or code can be written and uploaded into the equipment to produce the new products. Programmable automations are also characterized by high investment for the generic equipment, low production rates relative to fixed automation but the ease of flexibility with changes in product design and configuration [6]. The system is the most suitable for batch production where job variety is low and product volume ranges from medium to high. Examples of this automation are: steel rolling mills, paper mills, the textile industry, and many more.

Flexible Automation

This is an extension of programmable automation. It is used in Flexible Manu-facturing Systems (FMS) which is a computer-controlled manufacturing system. The system is capable of producing a variety of products with virtually no time lost for changeovers from one product to the next [6]. There is no production time lost while reprogramming the system and altering the physical setup (tooling, fixtures, and machine setting). Consequently, the system can produce various combinations and schedules of products though it requires high investment for a custom-engineered system. The features that distinguish flexible automation from programmable automation include the capacity to change part programs and the physical setup without loss of production time. It is typically utilized in job shops and batch processes where product varieties are high and job volumes range from medium to low. Such systems typically use multi-purpose CNC machines, Automated Guided Vehicles (AGV) and others.

Integrated Automation

This denotes the complete automation of a manufacturing system, with all processes involved in a computer-controlled, digitally processed, and coordinated system through digital information processing. It includes technologies such as computer-aided design and manufacturing, computer-aided process planning, computer numerical control machine tools, flexible machining systems, automated storage and retrieval systems, automated material handling systems (such as robots and automated cranes and conveyors), and computerized scheduling and production control.

PROGRAMMABLE LOGIC CONTROLLER (PLC) PROGRAMMING

PLC is a solid-state control device or computerized industrial controller that performs discrete or sequential logic in an automation environment. According to 'National Electrical Manufacture Association [NEMA]', PLC is defined as a digital electronic device that uses programmable memory to store instructions and implement specific functions such as programming logic, sequence, timing, counting, and arithmetic operations to control electronic machines and technical processes [9].

The basic components of PLC are the input and output (I/O) modules, power supply, control processing unit (CPU), memory system, communication protocols, and programming. The basic components of PLC are shown in Fig. (**9**).

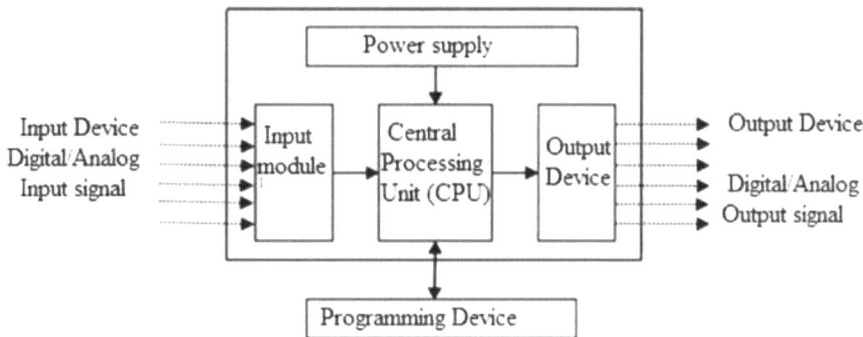

Fig. (9). Basic components of PLC.

The Input and Output (I/O) Modules

This is part of the PLC that connects the brain of the PLC, the CPU, to the machines. In a PLC system, there are modules usually dedicated to inputs and outputs. The module which interacts with the input signal is called input module. It is required to connect input devices like different types of switches. The module that interacts with the output signal is called output module. This is needed to connect output devices like electric applications. The classification of I/O modules of PLC is based on the two types of signals; discrete and continuous signals. Regarding classification based on the signals, I/O modules are classified further into two main parts; digital I/O module and analogue I/O module. Fig. (**10**) shows the classifications of the I/O module of the PLC.

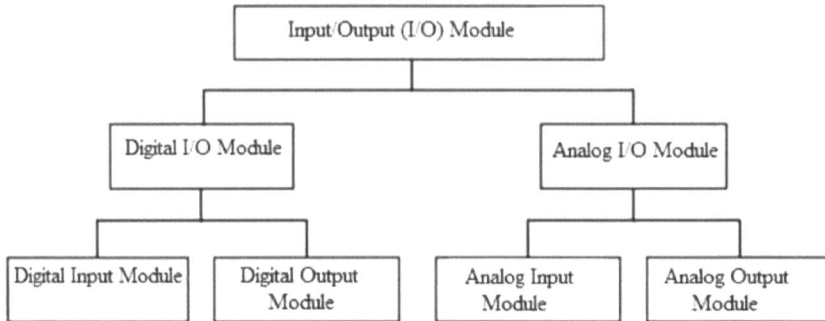

Fig. (10). Classification of input and output (I/O) modules of PLC.

In this section, analogue input and analogue output modules are discussed.

ANALOGUE INPUTS AND ANALOGUE OUTPUT

Analogue signals are universal signals in nature. They are representations of numerical quantities by the measurement of continuous physical variables in contrast to binary/ discrete signals which can only assume two states (0 or 1, ON and OFF). Some processes require analogue signals within a defined range for their operation. Examples of such signals include temperature, pressure, and flow just to mention a few. In order to understand the concept of analogue signal processing in Programmable Logic Controllers (PLCs), a clear emphasis on analogue-to-digital conversion (for analogue inputs) and digital-to-analogue conversion (for analogue output) must be declared.

Analogue Input Interface

In PLCs, analogue-to-digital conversion (continuous/real time signals to discrete time signal) is incorporated in analogue modules for analogue input processing. These modules sometimes are usually in-built in the controller, otherwise, cards/slots are usually integrated with the controller.

Analogue inputs can either be unipolar or bipolar depending on the signal received. Unipolar signals are signals with the possibility of receiving only positive input (voltage/current *e.g.,* 0-10V, 4-20mA, 0-5V, and 0-20mA). Bipolar signals are signals with a possibility of receiving both positive and negative (+/-) input (voltage/current *e.g.,* +/-10V, and +/- 20V).

In PLCs, processing analogue signals proves very crucial as raw data (temperature, pressure, and so on) from the sensor needs to be converted into digital information. Although this is done *via* the analogue digital (A/D) converter in the analogue module hence, manipulation of these analogue signals requires the concept known as scaling.

Scaling majorly involves the translation of these analogue values measured from the sensors/transducers (through voltage/current) within a defined range into digital information which the PLC understands [10]. Such a signal is then manipulated from a digital signal into a defined voltage/current for control of the analogue output. In order to idealize the concept of scaling for specifying analogue input, it is important to take into consideration three (3) major parameters which should be clearly defined for proper manipulation of the signal. These parameters include;

i. Type of analogue signal received (unipolar or bipolar): The signal type should be specified as either unipolar or bipolar signal type, depending on the manner in which the signal is perceived from the sensor, which helps in parameterization on the PLC. Hence, a bipolar input can be used to measure a unipolar signal thus it is not economical.

ii. The input range of the signal: The input range from the sensor output should be specified for easy parameterization on the controller. A voltage/current output given the defined range from the sensor should be known.

iii. Resolution: This is the smallest visible measurement increment in regard to the A/D converter's precision. When dealing with the resolution parameter, it is important to understand the distinct difference between bit resolution and voltage/current resolution and their relationship in analogue signal processing.

Bit resolution is often referred to as the number of bits the converter outputs. Every analogue module has its bit resolution for its conversion (usually 16-bit for Simatic products). However, conventional analogue modules are usually 11bits or 8bits. The 2^n (where n is the number of bits) gives the number of steps for a given bit resolution. Hence, for example, given an analogue converter *e.g.* a 12-bit converter, it would mean that there are 4096 individual ranges/steps.

Voltage/current resolution: If the number of steps for the 12-bit converter is 4096 with an input range of (0-10 v), then, it implies that for 0 - to 11-bit, each step would have a step voltage of 2.44mV *i.e.* 10/4096 = 2.44mV per voltage step of the analogue to digital converter. A more visible application can be seen in a temperature sensor using a range (0-100 °C) with a 12-bit resolution. The number of counts per degree in relation to the bit resolution of such converter = 4095/100

= 4.095 and the total voltage per step degree Celsius = 4.095 x 2.4mV = 9.8mV. This means that an output of 9.8mV from the sensor is sent into the analogue module.

Analogue Output Interface

This is similar to the analogue input interface but with the use of a digital to analogue converter for controlling the analogue output inbuilt into the module and un-scaling the converted signal from the input sensor for control. A practical example of analogue input and output in PLC programming is motor speed monitoring [11].

The Variable Frequency Drive (VFD) is connected to the analogue input with an address of IW0 (an input word in Siemens PLC) and an address name speed where the VFD feeds the analogue input real data within the input range of (+/- 50) rpm from the sensor. The required speed is monitored within the range and anything above shuts the motor off. The conversion with a 16-bit resolution and a (0-10V) is then done within the block to control the analogue output which the motor is connected *via* an analogue output card.

PLC PROGRAMING HIGHER FUNCTION

PLCs are simple to program, very flexible and can accommodate changes quickly without hardware modifications to the controller. They can also perform multiple functions. PLCs are reusable, reliable and designed for the industrial environment with ease of maintainability. The most common PLC programming languages include [12].

1. Ladder diagram (LD)

2. Sequential function charts (SFC)

3. Function block diagram (FBD)

4. Structured text (ST)

5. Instruction list (IL)

2-STEP CONTROLLER

This controller is also known as two-step, two-position, and ON and OFF [1-5]. It can be called a bang-bang controller also and conceptually, it is identical to proportional control with an incredibly high controller gain [2, 3]. It is common for simple, and uncritical applications. This controller is fundamentally a switch that is activated by the error signal and supplies just an on-off or high-low correcting signal as shown in Fig. (11) [1]. This kind of controller has only two positions: fully open or fully closed and it does not allow any in-between position [4]. Fig. (13) displays the response of a forward acting 2-step controller to a saw tooth error (see Fig. (12)) [3].

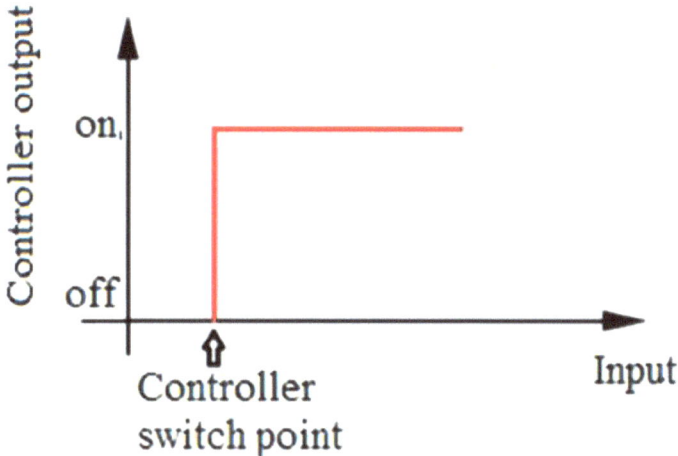

Fig. (11). 2-step controller.

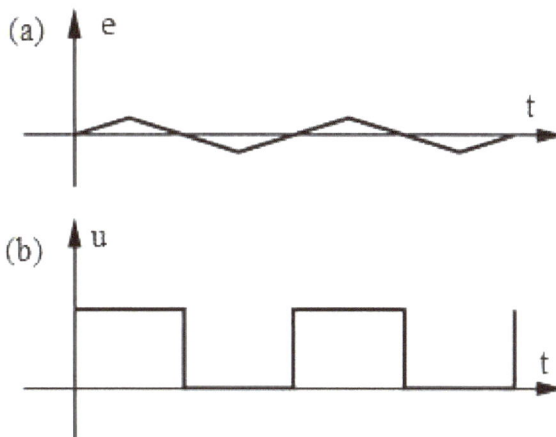

Fig. (12). Response of 2-step controller to sawtooth error.

Moreover, the 2-step controller is an automatic control and it can be utilized to control processes (like temperature processes) with high rise time [6].

The controller could be continuous or discontinuous. The output of the continuous controller varied continuously independent of the input unlike the output of discontinuous controllers that have only few defined values. The discontinuous controllers are the different groups of nonlinear controllers and static characteristics only define their dependence between input and output signals [6]. In practice, On/off controllers are the familiar examples of discontinuous controllers and their output $U(t)$ has two states only, that is, ON and OFF; HIGH and LOW [1, 2, 6] and as a result they are called two-step controllers at times. The output $U(t)$ of the 2-step controller can be defined as [2].

$$U(t) = \begin{cases} U_1 & \forall e(t) > 0 \\ U_2 & \forall e(t) \le 0 \end{cases} \tag{8}$$

Where: U_1 and U_2 are constants *On* and *off* values, *High* and *Low* values, *etc.*

Regulation process is achieved by switching between these two states and the change is instant, when process variable equals set point $r(t)$. As a result, oscillations of the $y(t)$ around its $r(t)$ occurs due to this working cycle as shown in Fig. (**13**). Moreover, there is a possibility that the controller would swiftly switch on and off many times as the $y(t)$ rises and meets $r(t)$. This may occur due to incorrect measurement, environmental interference, the noise and others.

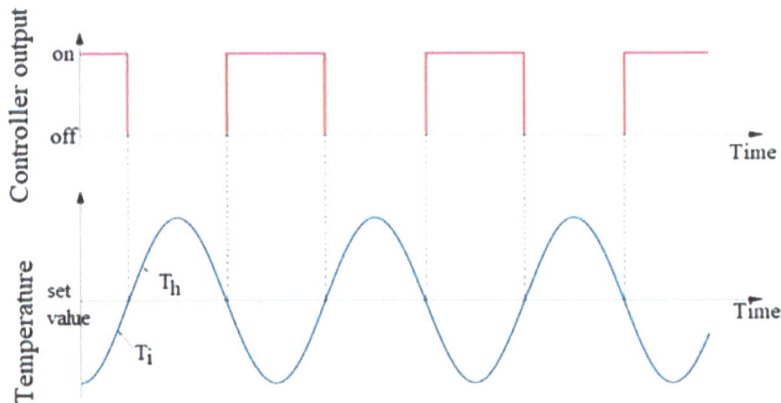

Fig. (13). Fluctuation of temperature about set value: T_i = Temperature below set value hence controller presents on signal, and T_h = Temperature continues to rise after controller off.

The cycling around the set point value is termed "hunting". However, rapid switching is not needed because it can cause lifetime reduction of a system or device being controlled or destroy the controller itself. With hysteresis, this problem could be eliminated [1]. It should be noted that apart from hysteresis, dead zone (dead band or neutral zone) can also be used to diminish the frequency of operation and wear on the components [1, 2, 5].

Characteristics of Hysteresis in a 2-Step Controller

The characteristics of hysteresis are as follows [6]:

1. It eliminates unwanted step changes.

2. It diminishes the precision of a regulation process.

3. It increases the amplitude of oscillations of the process variable.

4. It can be made by including a magnet with a pre-stressed spring or steel plate.

Advantages of a 2-Step Controller

The following are the advantages of 2-step controller [1, 2, 5, 6]:

1. It is economical.

2. It is simple.

3. Operation reliability.

4. With a system that has very large capacitance, it could maintain a stable value of the variable.

5. It can be implemented by mechanical switches (*e.g.* bimetallic strips or relays).

6. It is usually utilized in both domestic and industrial applications.

7. It is used in uncritical industrial applications that include some level control loops and heating systems.

Disadvantages of a 2-Step Controller

The following are the disadvantages of a 2-step controller [1, 2, 5, 7]:

1. Unstable output.

2. Due to its control action, there are time lags in the system and oscillations of the controlled variable take place about the desired condition.

3. A slight draught might activate it when it is at its set value.

4. High frequency of operation could cause wear on the components (such as other final control elements or control valves) if not properly taken care of.

5. It is sensitive to noise.

6. Less commonly employed than PID controllers since they are not as effective or as versatile.

BASICS IN CLOSED LOOP CONTROL

Closed loop control is designed to accomplish and maintain the required process condition by comparing it with the required condition, the set point value, to get an error value [5]. Fig. (14) shows the common form of a basic closed loop system.

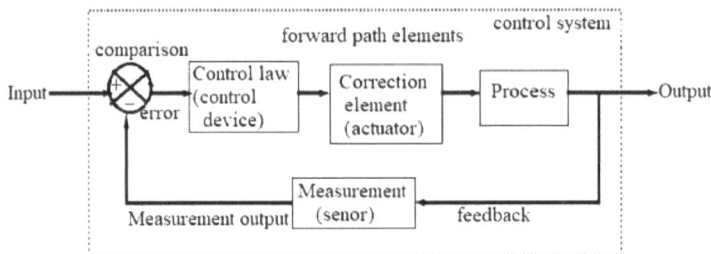

Fig. (14). Basic elements of a closed loop control system.

Comparison Element

This element compares the desired or required value of the variable being controlled with the measured value of what is being realized and generates an error signal $e(t)$ that can be defined as:

$$e(t) = u(t) - y(t)$$

(9)

Where: $u(t)$ and $y(t)$ are required value and measured actual value signals respectively. The value of $e(t)$ could be less than zero, equal to zero, or more than zero. When, $e(t) = 0$ there is no error then the output is the desired value and no signal produced to activate control. However, there is error if $e(t) \neq 0$ then there will be an error signal and control action will be activated.

Control Law Implementation Element

This element decides the kind of action to be taken once an error signal is received. The control law or mode utilized by the element may be on/off, proportional, integral or derivative laws [1] and they could be either combined or used separately [5]. In case of on/off, the element will just supply a signal that will switch on/off when there is an error. Regarding proportional law, a signal that is proportional to the error size is supplied so that a small control signal is generated if the error is small and *vice versa* if the error is large. The control signal persists to increase as long as there is an error when the integral mode is considered and the control signal is proportional to the rate at which the error is changing when derivative mode is employed.

The term controller or control unit is used regularly for the combination of the comparison element (the error detector, and the control law implementation element) [1, 9]. A differential amplifier is an example of such an element and it has two inputs: one for set value and the second for the feedback signal. Any discrepancy between the two inputs is amplified to provide the error signal. There would be no resulting error signal when there is no discrepancy.

Correction Element

This element is also known as the final control element. It creates a change in the process which intends to change or correct the controlled condition. Actuator is the term utilized for the element of a correction unit that offers the power to carry out the control action. The directional control valves, electric motors and others are examples of correction elements. Directional control valves are employed to switch the fluid low direction and so control the movement of an actuator such as the movement of a piston in a cylinder while electric motors are considered to control the rotational speed of the motor shaft.

Process

It is the system wherein there is a variable that is being controlled. For instance, it could be a room within a house with the variable of its temperature being controlled. Sometimes, the process is regarded as a plant.

Measurement Element

This is the element that generates a signal related to the variable condition of the process that is being controlled. This could be a temperature sensor with appropriate signal processing.

Fig. (**14**) contains two major paths through the system taken by signals. These paths include forward and feedback paths.

Forward Path

This path is utilized for the path from the error signal to the output. The elements of this path include the control law, the correction, and the process elements as in Fig. (**15**).

Feedback Path

This path is a way whereby a signal related to the actual condition being accomplished is fed back to adjust the input signal to a process. When the signal that is fed back is deducted from the input value, the feedback is said to be negative, otherwise it is said to be positive. It is the negative feedback that is needed to control a system.

Frequently, the term *process control* is being employed to explain the control of variables. For example, the flow of fluids or liquid level, associated with a process so as to maintain them at some values. It should be noted too that at times the term *regulator* is used for a control system for maintaining a plant output constant in the presence of external disturbances. Thus, the term regulator is at times applied to the correction unit.

Temperature, flow rate, pressure, liquid level, and composition are the major groups of measurements utilized in process control [2].

The performance of different process parameters, such as temperature, pressure, flow, and level, is necessary to be monitored and controlled in the process industries to achieve desired production rate, worker safety, process stability, and many more. In industrial processes, there will be a need to control any of these parameters with regards to the fluid (*e.g.* water) in one, two, or more tanks, vessels, or reservoirs in order to prevent underflow or overflow and produce desired product. The closed loop control of each of these parameters is considered as follows:

Closed Loop Temperature Control

In a closed loop control, the input to the heating process relies on the difference between the desired temperature originally set (set point) and the real temperature fed back from the output of the system [1]. This difference is referred to as error.

Process controllers are the control system elements that modify the system output using the error signal as the input. On/off controller is the simplest controller but due to its limitations, PID or three-term controller that is capable of providing acceptable control in different circumstances can be used [1]. Proportional (P), integral (I) and derivative (D) are the three basic control laws or modes while the three term controller is the combination of all three laws.

For closed loop temperature control in process control, the temperature must be measured with the aid of a sensor in order to establish the error to be passed to the controller to supply the required input. Resistance temperature detectors (RTDs), thermocouples, and pyrometers are the common temperature sensors [2]. With the transmitter, the sensor output can be converted to the signal level suitable for controller input. The transmitters are designed to be direct acting in general and a temperature transmitter can be set for the sensor it was designed for. Thus, for a specific transmitter (temperature inclusive), the relationship between the transducer output and input can be defined as:

$$T_{out} = K_m \left(T - T_{min} \right) + R_{min} \tag{10a}$$

And

$$K_m = \left(R_{max} - R_{min} \right) / \left(T_{max} - T_{min} \right) \tag{10b}$$

where, T, T_{max} and, T_{min}, R_{max} and R_{min}, and T_{out} are temperature measured by desired temperature sensor, instrument span or range (maximum and minimum temperatures that the instrument is designed to measure), desired standard electrical signal range (that include: 0 to 10V, - 10 to 10V, 0 to 20mA, 4 to 20mA, and others), and transmitter output. Besides, K_m is a constant and it is equivalent to the gain of the measurement element as stated in [2].

Here, the temperature of the fluid tank shown in Fig. (**15**) is controlled by regulating the incoming fluid temperature. The in and out flow rates of this tank are steady.

Fig. (15). Tank temperature control.

Regarding this tank, the transfer function for the temperature control problem is defined by a first-order equation given as [10]:

$$G(s) = \frac{T_t(s)}{T_c(s)} = \frac{e^{-s\tau}}{a^{-1}s+1} \ , a = \dot{m}/M \tag{11}$$

or as [11]

$$G(s) = \frac{T_t(s)}{T_c(s)} = \frac{Ke^{-s\tau}}{\alpha s+1} \tag{12}$$

Where T, T, τ, $\dot{m}$, M, K and α are tank temperature, temperature of the supplied fluid, time delay for material transport in the pipe, mass flow rate ($\dot{m}_{in} = \dot{m}_{out}$), mass of the fluid contained in the tank, static gain, and inertia time constant respectively. If the supplied fluid is at the constant pressure, then, the time delay from the material transport will also be constant and *vice versa*. At this point, four different control laws were applied to keep the fluid in the tank at a desired temperature (set point = °F) using Eq. (11) where in $\alpha = 1$, and $\tau = 10$.

The responses obtained from these control laws are presented in Fig. (16). It can be observed in the figure that all the responses are stable and prevent overshoots that can affect the quality of the product and none of them oscillate unnecessarily. However, unlike others, PID was very fast to approach the desired set point.

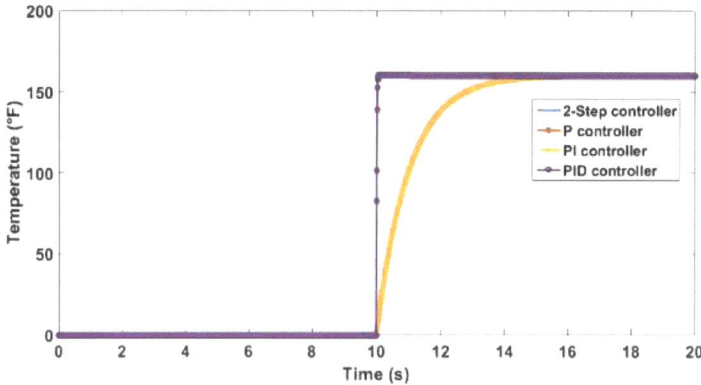

Fig. (16). 2-step,*P, PI*, and *PID* comparison for temperature control.

Closed Loop Level Control

The level of the fluid in the tank is measured by a sensor and it is adjusted by fluid pump and one outlet valve as shown in Fig. (**17**). It is assumed that there is no stiction of the valve that normally causes oscillations in close-loop systems. Hence, the valve position can attain its steady value under a constant reference if there is no valve stiction and *vice-versa* [12]. The pump motor speed and fluid level are varying with a control action. The speed increases right away to raise the tank level when it is reduced and *vice versa* [13].

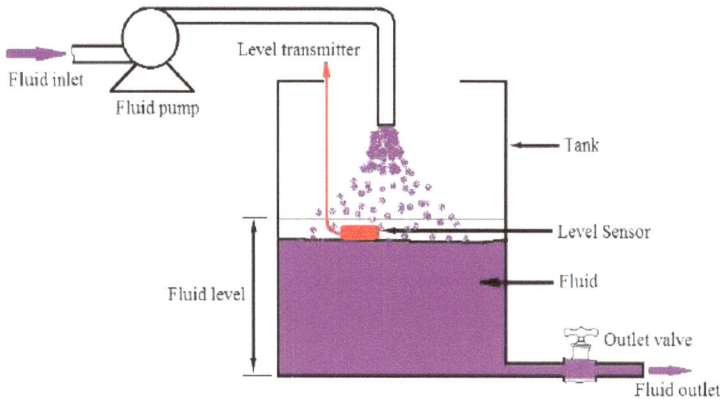

Fig. (17). Water tank system.

The fluid level in the tank can be measured by several instruments [1]: floats, differential pressure, displacer gauge, electrical conductivity level indicator, load cell, nucleonic level indicators, capacitive level indicator, and ultrasonic level gauge.

The fluid level sensor is located based on the required fluid level as in Fig. (**17**). As the fluid level in the tank is decreasing and reaches a certain level, the sensor will quickly send a signal to the controller. Consequently, controller will switch the pump ON in order to re-fill the tank. As time passes by, the sensor will send a signal back to controller again to switch OFF the pump when the tank is filled with the fluid to its desired level. This scenario will continue and it can be represented in block diagram as demonstrated in Fig. (**18**).

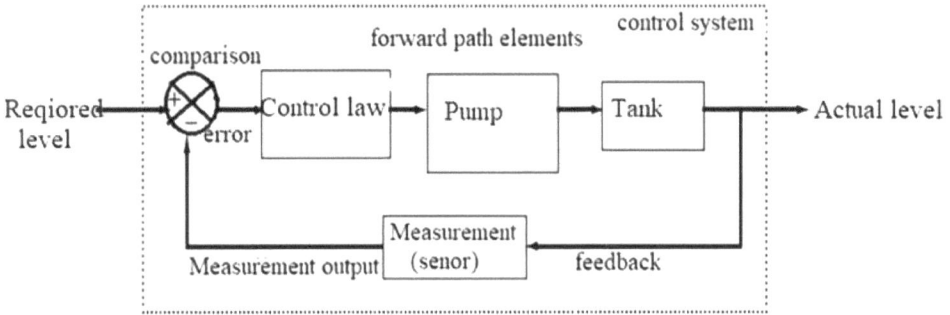

Fig. (18). The block diagram to represent the water tank scenario in Fig. **17**.

For steady fluid flow, the differential equation for the system in Fig. (**18**) with a constant value of *R* can be obtained as [14-16].

$$CR\frac{dh}{dt} = Rq_i - h \,. \tag{13}$$

Then, with q_{in} and *h* are input and output respectively, the transfer function *G*(*s*) of Eq. (13) is obtained as:

$$G(s) = \frac{H(s)}{Q_{in}(s)} = \frac{D}{\tau s + 1}, \quad \tau = CR = AD \,. \tag{14}$$

or

$$G(s) = \frac{H(s)}{Q_{in}(s)} = \frac{1}{As + D^*} \tag{15}$$

Where *h* is the small deviation of head from its steady-state value *(m)*, q_{in} is the small deviation of inflow rate from its steady-state value (m^3 / s), q_{out} is the small deviation of outflow rate from its steady-state value (m^3 / s), *C* is the capacitance of the tank (m^2), A is the cross-sectional area of the tank (m^2), R is the resistance for liquid flow, D is the slope of the curve at the operating point, CR or AD is the time constant of the system. By comparing Eqs. (14) and (15), it would be observed that $D^* = 1 / D$.

Besides, there is relationship between the: pump voltage and flow rate, and tank liquid level and sensor measurement. Thus, If the input flow rate $Q_{in}(s)$ is designed in such a way that it can be regulated by adjusting the applied voltage $V_{in}(s)$ to the pump motor ($Q_{in} = K_p V_{in}$) and the liquid level sensed by a transducer generates output voltage $V_s(s)$ that is proportional to the liquid level ($V_s = K_s H$); then, Eq. (14) can be re-expressed as:

$$G(s) = \frac{V_s(s)}{V_{in}(s)} = \frac{K}{\tau s + 1}, \ K = DK_s K_p \tag{16}$$

Where K_p, K_s, and K are gains.

The height and diameter of the tank considered at this juncture are presumed to be $6m$ and $2m$ in that order and the maximum level of the fluid in the tank is expected to be $5m$. The time constant $\tau = CR = AD$ or τ is known to be 63.2% of the final value, then:

$$\tau = 63.2\% \times 5 = 3.16\,s \tag{17}$$

And

$$D = \tau / A = 1.0059\,m^2 .$$
$$\text{Also, if } Q_{in} = Q_{out} \text{ and } Q_{out} = R^{-1}H(s), \text{ then} \tag{18}$$

$$Q_{out} = 4.9707 \approx 5\,m^2 / s . \tag{19}$$

In addition, let the maximum voltage V_{in} to be sent to the pump motor is 20V and the maximum Q_{in} is as obtained in Eq. (19), then by using $Q_{in} = K_p V_{in}$, $K_p = 0.1 m^3 / sV$. Similarly, by using standard electrical signal range V_s (e.g 0 to 10V with desired minimum and maximum fluid levels are 0 and $5m$ as stated earlier) for transmitter, K_s can be obtained from $V_s = K_s H$ to be $2V / m$. Thus,

$$K_p = 0.1 m^3 / sV , \text{ and } K_s = 2V / m \tag{20}$$

By using Eqs. (18) and (20) in Eq. (16), it yields:

$$K = 0.2012\,m^4/s \tag{21}$$

Now, by substituting required among Eqs. (17) to (21) in Eqs. (14) to (16), they become:

$$G(s) = \frac{H(s)}{Q_{in}(s)} = \frac{D}{\tau s + 1} = \frac{1.0059}{3.16s + 1}, \tag{22}$$

$$G(s) = \frac{H(s)}{Q_{in}(s)} = \frac{1}{As + D^*} = \frac{1}{\pi s + 0.9941}, \tag{23}$$

And

$$G(s) = \frac{V_s(s)}{V_{in}(s)} = \frac{K}{\tau s + 1} = \frac{0.2012}{3.16s + 1}. \tag{24}$$

Four different control laws were considered in this section to keep the fluid in the tank at a desired level (set point=5m). Eq. (22) only was considered for level control. The responses obtained from the selected control laws are as shown in Fig. (**19**).

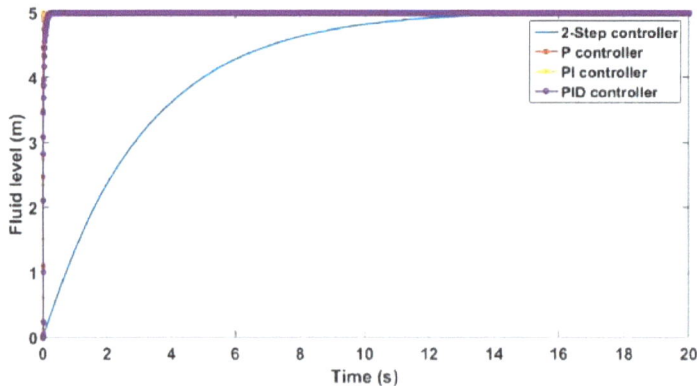

Fig. (19). 2-step, *P, PI*, and *PID* comparison for level control.

It can be observed in Fig. (**19**) that all the responses are stable and prevent overshoots that can affect the quality of the product and none of them oscillate unnecessarily. However, unlike the 2-step controller, other controllers were very fast to approach the desired set point. Although the difference between P, PI and PID was not significant as it can be noticed in the figure, the PID also performed better than P and PI while PI performed better than P.

Closed Loop Flow Control

In this subsection the fluid flow rate is expected to be controlled instead of fluid level in the tank shown in Fig. (**17**). The set point needs to be created to control the flow rate of the fluid flowing through the pipes of the system [17]. Some of the instruments that can be employed to measure fluid flow include: turbine flowmeter, differential pressure methods, coriolis flow meter, vortex flow rate method, and ultrasonic time of flight flow meter [1]. These instruments can also be grouped into: drag effects, obstruction meters, and magnetic flow meter [18]. The appropriate flow sensor and transmitter are needed for effective control of the fluid flow. The required flow control loop is similar to the one in Fig. (**19**) but with the flow sensor and transmitter.

For steady fluid flow, the differential equation obtained for the system in Fig. (**18**) with a constant value of R can be used here with modification for the flow control. As a result, with q_{in} and q_{out} are input and output flow rates respectively, the transfer function of the equation can be expressed by substituting $H(s) = RQ_{out}(s)$ in Eq. (14) as:

$$G(s) = \frac{Q_{out}(s)}{Q_{in}(s)} = \frac{1}{CRs+1} = \frac{1}{ADs+1} = \frac{1}{\tau s+1}, \quad \tau = CR = AD. \tag{25}$$

Similarly, since there is relationship between the pump voltage and flow rate, if the input flow rate $Q_{in}(s)$ is designed in such a way that it can be regulated by adjusting the applied voltage $V_{in}(s)$ to the pump motor $(Q_{in} = K_p V_{in})$, then, Eq. (25) can be re-defined as:

$$G(s) = \frac{Q_{out}(s)}{V_{in}(s)} = \frac{K_p}{\tau s+1} \tag{26}$$

Where K_p is the gain.

Now by substituting Eqs. (17) and (20) in Eqs. (25) and (26), they become

$$G(s) = \frac{Q_{out}(s)}{Q_{in}(s)} = \frac{1}{\tau s+1} = \frac{1}{3.16s+1} \tag{27}$$

And

$$G(s) = \frac{Q_{out}(s)}{V_{in}(s)} = \frac{K_p}{\tau s + 1} = \frac{0.1}{3.16s + 1} . \qquad (28)$$

Also, different control laws used in previous sections were considered here to keep the fluid flow at a desired flow (set point ∼ $5m^3 / s$). Eq. (27) only was employed for the fluid flow control. The responses obtained from the selected control laws are as shown in Fig. (20).

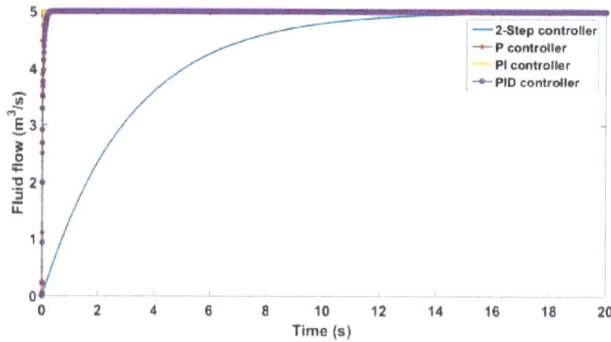

Fig. (20). 2-step, *P, PI*, and *PID* comparison for flow control.

It can be observed in Fig. (21) that all the responses are stable and prevent overshoots that can affect the quality of the product and none of them oscillate unnecessarily. In addition, the performance of all controllers employed similar to what was observed under fluid level control.

In general, with few exceptions, there are no stability problems in flow control loops and as a result the PI controller was considered for almost all flow control loops [5, 9]. Derivative action is regularly not considered in flow control to prevent noise that may lead to unacceptable control [9].

Closed Loop Pressure Control

Several sensors are available in industry to monitor fluid pressure in industrial processes. In industrial processes, fluid pressure monitoring entails the monitoring of the elastic deformation of bellows, diaphragms, and tubes. Some of the familiar sensors used in industrial processes include [1]: piezoelectric sensor, diaphragm sensor, and bourdon tube. Fluid pressure control is just like fluid flow control, the pressure at the beginning of a pipeline is related directly to flow in the line, and inertia of the fluid flowing is the only process dynamic contribution [7].

In industrial control systems, low pressure pneumatic controllers are employed in industrial processes to a large extent. This may be due to their simplicity, effortlessness of maintenance, and explosion proof character [14]. Several industrial processes and pneumatic controllers entail the flow of air or a gas through joined pipelines and pressure vessels as shown in Fig. (**21**).

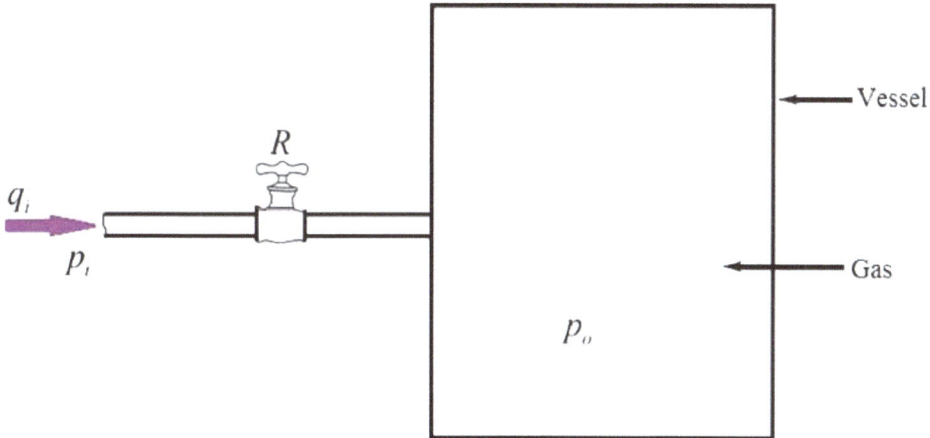

Fig. (21). A pressure system.

The system shown in Fig. (**21**) can be seen as linear if there are small deviations only in the variables from their corresponding steady-state values. Consequently, by using law of conservation of mass, the first other differential equation for the system can be obtained as [14, 19].

$$CR\frac{dp_o}{dt} = p_i - p_o \tag{29}$$

where C, R, p_i and p_o are capacitance, resistance, small change in inflow gas pressure, and small change in gas pressure in the vessel. Eq. (29) is the mathematical model of the pneumatic system going through a polytropic process with an ideal gas. In addition, if the value of the polytropic exponent in C is selected to be 0, 1, ∞, or heat capacity ratio; Eq. (29) remains valid for an isochoric, isothermal, isobaric, or isentropic process [19].

With p_i and p_o are the input and output in that other, the transfer function $G(s)$ for the system described by Eq. (29) can be determined as.

$$\frac{P_o(s)}{P_i(s)} = \frac{1}{\tau s + 1}, \; \tau = CR. \tag{30}$$

As in previous sections, all the used control laws were considered here to keep the pressure in the vessel at a desired pressure (30 KN / m^2). Eq. (30) with and $R = 0.167$ x $10^{10} N\, kg / m^2$ s and $C = 5.75$ x $10^{-9} N\, kg / m^2$ was employed for air pressure control. The responses obtained from the selected control laws are as shown in Fig. (**22**).

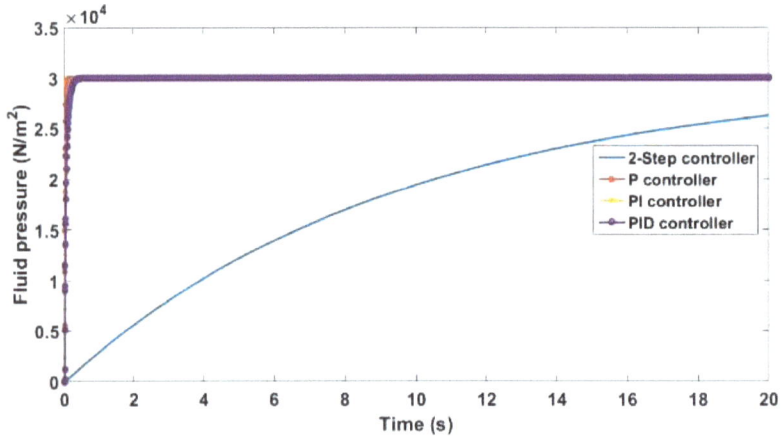

Fig. (22). 2-step, *P, PI, PID* and comparison at desired flow.

It can be observed in Fig. (**22**) that all the responses are stable and prevent overshoots that can affect the quality of the product and none of them oscillate unnecessarily. The performance of each controller was similar to its performance in the previous section.

CONCLUSION

Different control laws that can be used to control industrial processes are considered. It was observed that in all the processes considered, all the responses are stable, no overshoot that can affect the quality of the product, and none of them displayed needless oscillation. However, while other controllers approached the desired set point very fast, the reverse was the case regarding the 2-step controller. In all, there was no significant difference between P, PI, and PID controllers' performances in all the processes except under temperature control where PID performed far better than P and PI controllers. Nevertheless, the PID controller displayed the best performance in all the processes followed by PI and P controllers as it can be observed in all the figures for the processes considered. This shows that it will continue to be important in several applications in the Industries while PI and P will follow in that other.

REFERENCES

[1]   W. Bolton, *Instrumentation and control systems* Elsevier Science & Technology Books., 2004.

[2]   D.E. Seborg, T.F. Edgar, and D.A. Mellichamp, *Process Dynamics and Control.* John Willey & Sons, Inc.: USA, 2004.

[3]   J. Love, *Process Automation Handbook:A Guide to Theory and Practice.* Springer-Verlag London Limited: London, 2007.

[4]   S. Faizollahzadeh Ardabili, A. Mahmoudi, T. Mesri Gundoshmian, and A. Roshanianfard, "Modeling and comparison of fuzzy and on/off controller in a mushroom growing hall", *Measurement,* vol. 90, pp. 127-134, 2016.
       [http://dx.doi.org/10.1016/j.measurement.2016.04.050]

[5]   W. Altmann, *Practical Process Control for Engineers and Technicians.* Newnes: London, 2005.

[6]   T. Uriča, and A. Simonová, "Simulation of an on-off controller for systems of second order with the use of labview", *Procedia Eng.,* vol. 192, pp. 905-910, 2017.
       [http://dx.doi.org/10.1016/j.proeng.2017.06.156]

[7]   F.G. Shinskey, *Process Control Systems:Application, Design, and Adjustment.* McGraw-Hill: Toronto, 1988.

[8]   K. Warwick, *An Introduction to Control Systems.* World Scientific: Singapore, 1996.
       [http://dx.doi.org/10.1142/2175]

[9]   D.R. Coughanowr, *Process Systems Analysis and Control.* McGraw-Hill, Inc.: London, 1991.

[10]   G. Zhiqiang, T.A. Trautzsch, and J.G. Dawson, "A stable self-tuning fuzzy logic control system for industrial temperature regulation", *Proc. Conference Record of the 2000 IEEE Industry Applications Conference. Thirty-Fifth IAS Annual Meeting and World Conference on Industrial Applications of Electrical Energy,* vol. 1232, 2000pp. 1232-1240
       [http://dx.doi.org/10.1109/IAS.2000.881990]

[11]   M. Laskawski, and M. Wcislik, "Sampling rate impact on the tuning of pid controller parameters", *Int. J. Electron. Telecommun.,* vol. 62, no. 1, pp. 43-48, 2016.
       [http://dx.doi.org/10.1515/eletel-2016-0005]

[12]   L. Fang, J. Wang, and X. Tan, "Analysis and compensation of oscillations induced by control valve stiction", *IEEE/ASME Trans. Mechatron.,* vol. 21, no. 6, pp. 2773-2783, 2016.
       [http://dx.doi.org/10.1109/TMECH.2016.2559510]

[13]   V. Saravanan, and R.B. Kumar, "Mathematical modelling and controller design using electromagnetic techniques for sugar industry process", *Automatika,* vol. 62, no. 2, pp. 155-162, 2019.
       [http://dx.doi.org/10.1080/00051144.2019.1653667]

[14]   K. Ogata, *Modern Control Engineering.* Prentice Hall: New Jersey, 2002.

[15]   H. Klee, and R. Allen, *Simulation of dynamic systems with matlab® and simulink.* CRC Press: New York, 2011.

[16]   F.R. Betancourt, F.H. Escobar, and A.J.C. Esquivel, "Mathematical modeling and experimental identification of the ce 105mv tank system", *J. Eng. Appl. Sci.,* vol. 15, no. 19, pp. 2070-2078, 2020.

[17]   L.F. Alves, D. Brandão, and M.A. Oliveira, "A multi-process pilot plant as a didactical tool for the teaching of industrial processes control in electrical engineering course", *Int. J. Electr. Eng. Educ.,* vol. 56, no. 1, pp. 62-91, 2019.
       [http://dx.doi.org/10.1177/0020720918787455]

[18]   J.P. Holman, *Experimental Methods for Engineers.* McGraw-Hill Madrid: Madrid, 2001.

[19]   R.S. Esfandiari, and B. Lu, *Modeling and Analysis of Dynamic Systems.* CRC Press: New York, 2018.
       [http://dx.doi.org/10.1201/b22138]

CHAPTER 4

Levels of Automation

Ilesanmi Afolabi Daniyan[1,*], Lanre Daniyan[2], Adefemi Adeodu[3] and Ikenna Uchegbu[4]

[1] *Department of Industrial Engineering, Tshwane University of Technology, Pretoria0001, South Africa*

[2] *Department of Instrumentation, Centre for Basic Space Science, University of Nigeria, Nsukka, Nigeria*

[3] *Department of Mechanical Engineering, University of South Africa, Florida, South Africa*

[4] *Department of Mechanical & Mechatronic Engineering, Afe Babalola, Ado Ekiti, Nigeria*

Abstract: This chapter discusses the levels of automation (LOA). The degree to which a system, process or task is automated is referred to as the level of automation. They are: manual, semi-automatic and fully automatic depending on the level of human involvement, the system or processes to be automated and the end users' requirements. At the lowest level; the manual represents the human control level while the fully automatic level represents the computer controls level. At the semi-automatic level, the control activities involve both human and computer controls. The human control tasks include sensory processing for information acquisition, perception for information analysis, decision-making based on cognitive processing for action selection, and response selection for action implementation. Furthermore, this chapter also highlights the elements of system automation and classes of automated systems. The identification and specifications of the elements of the system's automation based on the end-user requirements are a critical aspect of the control design phase. The major elements of the system's automation include a sensor, a controller, an actuator, a power component, a motor and drives, a communication protocol, a human-machine interface, *etc.* Classes of automation systems could also be fixed, programmable, flexible, integrated, or cognitive automation depending on the need. The future of fully autonomous systems is exciting and promising although many industrial processes and systems are semi-autonomous thus relying on human factors such as physical, mental and technical capabilities such as intuition, perception, sensitivity, observation, experience, and judgment to arrive at effective decision making as it relates to system's control.

Keywords: Computer control, human control, LOA, system automation.

---

[*] **Corresponding author Ilesanmi Afolabi Daniyan:** Department of Industrial Engineering, Tshwane University of Technology, Pretoria 0001, South Africa; Tel: +27 (064) 5298778; E-mail: afolabiilesanmi@yahoo.com

INTRODUCTION

The level of automation in a system or process is the degree of human or computer controls employed for a particular operation which ranges from direct manual control to fully automated controls [1]. The levels of automation are defined as a continuum from a manual mode of operation to a fully automatic mode [2]. This refers to the number of manning levels that focus on the level of information sharing between machines and humans with varying degrees of human involvement [3, 4]. Levels of automation also refer to cognitive activities such as the system's ability to respond to changes and make decisions independently [5]. There are many activities that characterize process control, manufacturing and automation processes such as data collection, actuation, monitoring, decision-making, and control implementation amongst others. These can be performed independently or collaboratively by human or computer controls. Hoffman *et al.* [6] underscore the need for collaborative human or computer control during operation. This is due to the fact that both human and computer controls have distinctive merits and as such could compensate for the weaknesses inherent in either if employed collaboratively (Table **1**).

Table 1. Merits of the human and computer controls [6].

Human control	Computer (Machine) control
Inductive reasoning.	Deductive reasoning.
Ability to make decisions and exercise judgment.	Quick control and response to changes and signals.
Flexibility and ability to improvise.	Ability to perform routine and repetitive activities.
Effective perception of the work environment and working conditions such as light, sound, noise, temperature, *etc.*	Large information storage, processing and computational capacities.
Ability to detect a small amount of acoustic or visual energy.	Ability to multi-task and effectively handle complex operations.

Hence, Satchell [7] stated that man-machine integration can be a suitable for task sharing and control.

AUTOMATION LEVELS

Depending on the level of human intervention involved, the process of automation may be manual, semi-automatic, and fully automatic [3, 8-10]. A decision needs to be taken on the process of achieving vital production processes involving material handling, assembly, performance evaluation, trouble shooting, inspection and quality control feedback, data collection and decision making. These production processes can be achieved by manual, automated or semi-automated means. The automation process, however, depends largely on the size and type of industry, the nature of product, product complexity, environment, manufacturing structure, and the efficiency required amongst other factors [11, 12].

Manual Level

The manual level relies solely on human intelligence and effort to control, adjust or modify the operations of a system. It has the merits of flexibility, cost-effectiveness and is mostly suitable for small-scale industries with light equipment and simple operations or products that require a short production cycle. The demerits, however, lie in the low speed of production which increases production time, poor handling, and lower efficiency due to stress and fatigue when compared to automatic means and most times not suitable for production with heavy equipment and complex products as well as a product requiring reproducibility.

Semi-Automatic Level

This level utilizes the combination of both human and computer controls or artificial intelligence for control. Thus, it represents a flexible mix of manual and automated tasks which is relatively flexible to accommodate some changes during the course of production. Computer control systems offer sets or a restricted set of alternatives regarding process conditions that humans have to make decisions about and implement. Its overall efficiency and delivery are greater than the manual process but lower than the fully automatic process.

Fully Automatic Level

This level relies solely on computer algorithms and artificial intelligence for control. At this level, the control is executed solely by the computer. The control system acquires information, analyzes and displays the information acquired, and further decides on the necessary actions based on the outcome of the analysis and implements the actions based on the decision. The control system performs the tasks of data acquisition, process tracking, monitoring, provision of alternatives, monitoring, supervisory control and reporting. They are costly but highly efficient and can carry out simple as well as complex industrial operations with a high degree of precision and accuracy within a short time. The pace of production delivery with high precision will eventually offset the high initial cost. This process is inflexible to accommodate interchangeability or process deviation that is not programmed from the onset. Its lack of flexibility makes it catastrophic when there are challenges along production routes or in automatic mode. Fully automatic level boasts of high speed of production within a short production cycle, highly suitable for mass production, and efficient material handling amongst others.

Table **2** presents the involvement of human and computer controls in some selected activities for the three levels of automation.

Table 2. Human and computer controls in some selected activities for the three levels of automation.

Activities	Manual	Semi-automatic	Fully Automatic
Generation of alternatives	Human	Computer	Computer
Selection of alternatives	Human	Human	Computer
Implementation of alternative	Human	Human/Computer	Computer
Monitoring	Human	Human/Computer	Computer
Supervision control	Human	Computer	Computer
Action support	Human	Human	Computer
Processing	Human	Computer	Computer
Decision support	Human	Human/Computer	Computer

ELEMENTS OF SYSTEM AUTOMATION

The following are basic elements of an automated system namely:

Sensor

The sensor provides feedback to the controller. The controller detects and responds to changes with the aid of the sensor. The sensor measures the condition of the process as well as the output of the system while the transmitter converts the measured variable into an optical or electrical signal. It provides input from the process and from the external environment. A temperature probe in a reactor is an example of a sensor that senses heat.

Controller and I/O

The automatic controllers are a programmable controller that uses a software command to manipulate the system or adjusts a variable response to a command and the feedback system. The threshold value also known as the desired value or set point is set on the controller while the controller decides whether or not the process is acceptable. A programmable controller can be a Programmable Logic Controller (PLC), Programmable Automation Controller (PAC), or PC depending on the system's complexity or end-user specification or requirements. The discrete and analog input and output connect the controller to the system's sensor and actuators. The signal can originate from the main control panel *via* a terminal strip wired to a field device. However, the distributed I/O architecture can reduce the amount of wiring by multiplexing multiple I/O signals with the use of a single cable from the remote I/O component to the control panel.

Actuator

This is the component that responds to the input from the controller. It is a control element that executes instructions from the controller by giving the output of an action that is meant to change the control variable. The actuator also converts systems output into physical movements. Examples of actuators are: valves, pumps, speed drives, bakes, relays, solenoids, servo motors, *etc*.

Power Component

The power component gives power to other components such as motors, controllers, drives, *etc*.

Motors and Drives

The motors and drives are components of the system's control. The motor is a device that converts linear or rotational force to power the system while a drive is an electronic device that controls the electrical energy of the motor in terms of speed and torque. Motor control devices include starters, contractors, drives, *etc*.

Communication

The communication protocol of the distributed I/O communicates the data acquired from the sensor directly to the system controller. It also delivers the basic diagnostic and detailed status of the system to the controller. The communication sub-system also includes multiple Ethernet and serial port that integrates a variety of devices, Human-Machine Interface (HMI), *etc*. to facilitate effective networking. The modern system often uses USD, microSD communication and storage devices.

Multiple high-speed Ethernet ports ensure effective and responsive communication or networking. This links the webserver and remote access communication for effective system control.

Human-Machine Interface (HMI)

The HMI shows important information about the system's condition *via* the graphical or textual display. The HMI can be in the form of panel display Liquid Crystal Display (LCD), monitors, or touch panels. They are essential for control and monitoring as well as reporting the system's status. The identification of the end-user requirements at the design phase will determine the HMI capability and size.

The HMI can also be the data hub of the system by connecting to multiple networking devices. Where multiple protocols are used, the HMI can serve as the protocol conversion for data exchange among other smart devices. The HMI can also be used to send the data acquired to cloud storage to enable remote access *via* the Internet.

CLASSES OF AUTOMATED SYSTEMS

Automated systems generally fall into five categories namely;

1. Fixed automation

2. Programmable automation

3. Flexible automation

4. Integrated automation

5. Cognitive automation

Fixed Automation

This is a system in which the sequence of assembly operations or job processing is fixed by the virtue of equipment configuration. In other words, it uses mechanized machinery to perform fixed and repetitive operations in order to produce a high volume of similar parts. Under fixed automation, the system uses specific programs with the aid of pneumatic logic or numerical control to execute a sequence of operations without compensation for the error generated or deviation from the normal process. It is used in high-volume production with dedicated equipment, which has a rigid set of operations and is designed to be efficient for this set. It is predominantly used in continuous flow and discrete mass production systems. For example in the distillation process, conveyors, paint Shops, transfer lines, *etc*.

The features of the fixed automation system are as follows;

1. It is rigid to product varieties such that the assembly or production lines are inflexible to accommodate changes.

2. It is characterized by high initial investment and high production rates. Their optimal production rate within a short cycle time can offset the high initial cost of investment.

3. Low variability in product type in relation to size, shape, part count, and material.

4. Predictable and stable demand for a 2 to 5-year time period, so that the manufacturing capacity requirement is also stable.

5. It is characterized by high efficiency and low unit cost compared to others.

Programmable Automation

The assembly or production lines are designed to change with a sequence of operations in order to accommodate varieties of product configurations. The production line is controlled by a coded set of instructions known as a program which can be read and decoded by the system hence programmable automation is employed for a dynamic sequence of operations and system configuration using electronic controls. However, significant programming efforts are often required to reprogram the system or sequence of operations. Under programmable automation, the system uses a dynamic program to execute a sequence of operations with adequate compensation for the error generated or deviation from the normal process. Programmable automation finds application in steel rolling mills, paper mills, *etc*. The features of this type of manufacturing system are as follows;

1. It is flexible to changes in product varieties such that the assembly or production lines can change to accommodate other operations.

2. The production rates are lower when compared to the fixed automation system.

3. Investment in programmable equipment is less, as the production process is not changed frequently. It is typically used in batch processes where job variety is low and product volume ranges from medium to high, and sometimes in mass production.

4. Low unit cost for large batches.

5. High unit cost compared to fixed automation.

Flexible Automation

Flexible automation is an extension of the programmable automation used in Flexible Manufacturing Systems (FMS) in which a computer-controlled system performs different tasks by changing from one task or production line to another. Human operators are employed to give a high-level of commands in the form of codes into the computer systems identifying productproducts, operations or tasks as well as their location in the sequence and changes are done automatically. Each production machine receives settings or instructions from the computer system. These automatically load or unload the required tools and carry out their processing instructions. Hence, products are automatically transferred from one machine to the next in the sequence for processing. It is typically employed in industries with multiple production lines requiring frequent product changes as well as job shops and batch processes where product varieties are high and job volumes range from medium to low.

Flexible automation is characterized by the following features;

1. Significant variability in product type where the product mix requires a combination of different parts and products to be manufactured from the same production system.

2. Product life cycles are short. Frequent upgrading and design modifications alter production requirements.

3. Production volumes are moderate with varying demands.

4. High initial investment.

5. High unit cost compared to fixed or programmable automation.

Integrated Automation

It involves complete automation of the manufacturing system in which all the process operations function with the aid of computer control and are coordinated through digital information processing. Sometimes integrated automation is divided into simple modules to control subsystems in order to achieve a comprehensive overall system control. It includes advanced technologies such as computer-aided design and manufacturing, computer-aided scheduling and process planning, computer numerical control machines, flexible and integrated manufacturing systems, automated component or parts storage and retrieval systems, automated material handling and transportation systems such as robots and automated conveyors and cranes, computerized scheduling and production control. Integrated automation links the operating system through a common database. In other words, it involves the full integration of process and management operations using information and communication technologies. Typical examples of such technologies include; Advanced Process Automation

Systems and Computer Integrated Manufacturing (CIM).

Cognitive Automation

Cognitive automation is a self-learning emerging type of automation enabled by the simulation of human thought processes in a computerized model. Its primary concern is the automation of clerical tasks and workflows that consist of data that are unstructured. Cognitive automation depends on multiple fields such as natural language processing, real-time computing, machine learning algorithms, big data analytics, and evidence-based learning. It enables the replication of human tasks and a sense of judgment at high speeds and considerable scale by bringing intelligence into information processing and management with the use of the software. It offers the benefits of cost saving, low rates of error, improved business flexibility, and effective management of business processes in order to improve customer satisfaction.

It includes activities such as data extraction and synthesis, data reporting, contract management, natural language search, verification of manual activities and verifications, periodic follow-up and email communications.

CONCLUSION

The levels of automation are generally divided into three categories depending on the level of human involvement. The lowest level is manual, whereby controls and activities are accomplished without any computer control support or other forms of technology. The intermediate level is the semi-automatic level, which consists of a collaboration between the human and computer controls to achieve the required task. The highest level of physical automation is the automatic level, which excludes any form of human involvement from controlling and conducting physical tasks.

REFERENCES

[1]   J. Frohm, "Levels of automation in manufacturing", *International Journal of Ergonomics and Human Factors,* vol. 3, no. 3, pp. 1-28, 2008.

[2]   R. Parasuraman, T.B. Sheridan, and C.D. Wickens, "A model for types and levels of human interaction with automation", *IEEE Trans. Syst. Man Cybern. A Syst. Hum.,* vol. 30, no. 3, pp. 286-297, 2000. [http://dx.doi.org/10.1109/3468.844354] [PMID: 11760769]

[3]   M.P. Groover, *Automation, production systems, and computer-integrated manufacturing.* Prentice Hall: Upper Saddle River, N.J., USA, 2001.

[4]   P. Satchell, *Innovation and Automation.* Ashgate: London, England, 1998.

[5]   M.R. Endsley, E. Onal, and D.B. Kaber, "The Impact of intermed iate levels of automation on situation awareness and performance in dynamic control systems", *Proceedings of Proceedings of the 1997 IEEE 6th Conference on Human Factors and Power Plants,* 1997 Orlando, FL, USA

[6]   R.R. Hoffman, P.J. Feltovich, K.M. Ford, D.D. Woods, G. Klein, and A. Feltovich, "A rose by any other name. would probably be given an acronym", *IEEE Intell. Syst. Appli.,* vol. 17, no. 4, pp. 72-80, 2002.
[http://dx.doi.org/10.1109/MIS.2002.1024755]

[7]   H.A. Ruff, S. Narayanan, and M.H. Draper, "Human interaction with levels of automation and decision-aid fidelity in the supervisory control of multiple simulated unmanned air vehicle", *Presence,* vol. 11, no. 4, pp. 335-351, 2002.
[http://dx.doi.org/10.1162/105474602760204264]

[8]   R. Parasuraman, "Designing automation for human use: empirical studies and quantitative models", *Ergonomics,* vol. 43, no. 7, pp. 931-951, 2000.
[http://dx.doi.org/10.1080/001401300409125] [PMID: 10929828]

[9]   T.B. Sheridan, *Telerobotics, automation, and human supervisory control.* MIT Press: Cambridge, MA, USA, 1992.

[10]   T.B. Sheridan, *Humans and automation: System design and research issues.* John Wiley & Sons, Inc.: Santa Monica, USA, 2002.

[11]   M.R. Endsley, and D.B. Kaber, "Level of automation effects on performance, situation awareness and workload in a dynamic control task", *Ergonomics,* vol. 42, no. 3, pp. 462-492, 1999.
[http://dx.doi.org/10.1080/001401399185595] [PMID: 10048306]

[12]   T. Inagaki, "Situation-adaptive degree of automation for system safety", *Proceedings of 2nd IEEE International Workshop on Robot and Human Communication,* 1993 Tokyo, Japan
[http://dx.doi.org/10.1109/ROMAN.1993.367716]

CHAPTER 5

The Control System

Ilesanmi Afolabi Daniyan[1,*], Lanre Daniyan[2], Boitumelo Ramatsetse[3] and Khumbulani Mpofu[1]

[1] *Department of Industrial Engineering, Tshwane University of Technology, Pretoria 0001, South Africa*

[2] *Department of Instrumentation, Centre for Basic Space Science, University of Nigeria, Nsukka, Nigeria*

[3] *Department of Mechanical & Mechatronics Engineering, University of Stellenbosch, Stellenbosch, South Africa*

Abstract: This chapter presents the control system and its functions, types, examples, and representation of the process control systems. A control system is a system that regulates, directs, commands, and manages the performance of other sub-systems using a control loop. Basically, there are two major types of control systems, viz; the open and closed loop control systems. For the open loop control system, control action is independent of the desired output. This control system is referred to as a non-feedback control because of the absence of a feedback path. Although they are simple in design and relatively inexpensive but are less accurate compared to the closed-loop control system. On the other hand, for the closed-loop control system, control action is a function of the desired output. This control system is referred to as feedback control because of the presence of the feedback path. Although it is complex in design and expensive but more accurate compared to the open loop control system. To enhance the performance of basic controls, many modern systems incorporate advanced controls such as advanced regulatory controls, advanced process controls, multivariable predictive control, non-linear multivariable predictive control, fuzzy logic control, inferential measurements, *etc.* Advanced controls are a set of technologies employed to address a specific control deficiency in a system. While the basic controls facilitate the control of a system's basic operations, advanced controls are incorporated to enhance the performance of the basic controls.

Keywords: Basic control, Control system, Advanced control, Open loop system, Closed loop system.

INTRODUCTION

Production can be a challenging task amidst process or system variations, dynamic demand, and customer requirements if the production system is not

---

[*] **Corresponding author Ilesanmi Afolabi Daniyan:** Department of Industrial Engineering, Tshwane University of Technology, Pretoria 0001, South Africa; Tel: +27 (064) 5298778; E-mail: afolabiilesanmi@yahoo.com

effectively controlled. To deliver products with the right quality that will meet the production or demand requirements, the production system must be controlled to minimize errors and deviation. This chapter discusses the types of control systems and their functions, the types of system processes and the general control theory.

THE CONTROL SYSTEM

A control is a device used to manipulate the performance of a system. It is a term that can be used to describe the process of altering the behavior of a system or keeping the behavior constant over time. The alteration can be done either manually or automatically. This implies that controls can be done manually or automatically depending on the nature of the system. The system executes and monitors an industrial operation, taking corrective measures when there is a deviation from the set points. A control system can be defined as the collection of components that are designed to drive a given system from a particular input to the desired output. In other words, it is an interconnection of components including additional hardware that forms a system configuration to control the behavior of a dynamic system in order to obtain the desired system response.

FUNCTIONS OF THE CONTROL SYSTEM

The main function of a control system is to ensure that the outputs of a system do not deviate from the set or desired points [1, 2]. A good control system should effectively manage the instructions, direct and regulate the behaviour of a system or a system sub-component. Control systems are incorporated into the system for the following reasons [3]:

1. To obtain data (*via* direct measurement or acquisition): Data can be obtained *via* direct measurement of values from sensors which are read as input to process or provide signals as output. For instance, a sensor is a device for measuring physical quantities such as speed, temperature, pressure, velocity, *etc.* On the other hand, data can also be obtained *via* the acquisition of past activities.

2. To compare: This involves the dynamic comparison of the operations and conditions to the desired set points in real-time. This is to evaluate the measured value and the process value *vis-à-vis* the threshold. For instance, transducers are comparators that convert the non-electrical signal into an electrical value.

3. To compute: Control systems can be used to calculate the errors in a process or system. Error is the difference between the desired and true output signals. For instance, a transmitter can convert measurements from a sensor and send the signal to the main controller. The difference between the actual measurement and

the threshold value pre-set on the microcontroller represents the error generated.

4. To regulate: Systems are regulated, controlled, or corrected to alter the course of the operations or the process conditions to the desired set points. This is usually to curb the effect of error generated or external disturbances. They are also regulated to keep operating variables or output within the desired range. For instance, a controller provides logic for the process while the final control element also known as the actuator changes the process physically.

5. To track: Tracking is carried out to enable process variables or output to follow a particular changing form. It involves keeping the profiles of a series of operations and conditions.

Hence, the process of obtaining data, data comparison with the threshold, regulation of the course of operation, and tracking is what is referred to as the control actions of a control system.

TYPES OF CONTROL ACTIONS

There are two major types of control actions in a system namely discrete otherwise known as precise control and continuous control [2].

Discrete Control

The discrete control can be in the form of a simple binary ON/OFF control or a more advanced modulating control which is suitable when a sequential control of such a system is required. In discrete control, the system's variables are changed at discrete moments in time. The changes are implemented because of event-driven changes (changes in the state of the system) or because of time-driven changes (changes due to the fact that the specified amount of time has lapsed).

An event-driven change is usually executed by the controller as a response to a certain event that alters the state of the system. This change can be to initiate or terminate an operation, for instance, starting or shutting off a motor, opening and closing a valve, fuel tank used in a boiler-based power, the thermostat used on household appliances which either opens or closes an electrical contact plant, *etc.*

For a time-driven change, the control system implements control at a specified time or after a specified time has elapsed, for instance, heat treatment operations, automatic loading, and unloading of parts, the operation of a washing machine, *etc.*

A modulating control is an automated control, which is employed to regulate the amount of flow in the process. The control allows precise regulation of the flow rate *via* an actuator that uses the feedback and control signals to specifically open and close the system's valve. The feedback systems and Proportional-integral-derivative (PID) controllers are employed for precise control and continuous monitoring in these systems. The Proportional Integral Derivative (PID) controller finds application in the precise control of a system where a definite output control is desired.

The discrete control can be sequential in nature. Sequential control is done to make some effect occur in a system in a particular order which could be time-driven. It is one in which a programmed sequence of discrete operations is performed, often based on system logic that involves the system states. In sequential control, all the operations are carried out in a sequence to automate a system. The sequence of instructions is developed to switch the process or system between various statuses. It is an event-based control in which the sequence of activities needs to be controlled and not the variables. The sequence of occurrence of activities can trigger an action. Hence, one activity follows the other until the sequence is completed. For instance, an automated vehicle assembly line, and an elevator control system are examples of sequence control. The Programmable Logic Controller (PLC) is always employed for sequential control. Discrete control is usually employed in discrete manufacturing industries to control product outputs such as the number of parts or products, surface finish, dimensions and defects. Typical parameters for discrete control include force, acceleration, position and velocity, *etc.* and sensors such as photoelectric sensors, strain gauges, limit switches, piezoelectric sensors, *etc.* can be used for sensing the parameters.

Continuous Control

Continuous control requires the continuous monitoring of certain process parameters so that they remain at the set points or are adjusted to the desired set points. In this control, the monitoring of the output variable level against a set reference level will permit the identification of deviation (error) which can be amplified and used as an input (control) signal for control. Hence, in continuous control, the inputs are sent into the system continuously to control the output. Therefore, a change in the input signal will automatically lead to a change in the output signal.

The aim of this type of control is to achieve a desired set point or output signal *via* continuous monitoring of the output signal or adjustment of the control signal. Continuous control can compensate for perturbation without modifying the status of the process or system. Continuous control finds application mostly in the process industries for continuous processes and monitoring or control of outputs such as weight, liquid or solid volume, the concentration of the solution, confor-mity to specifications, *etc.* Typical parameters for continuous control include temperature, pressure, volume flow rates, and sensors such as thermocouples, pressure sensors, flow meters, *etc.* that can be employed for sensing the para-meters.

Typical examples of continuous control include:

1. The input of a chemical reaction, which is a function of temperature, pressure, flow, and rates of reactants.

2. The control of the workpiece position relative to the cutting tool in a milling operation.

The following are types of continuous control:

Regulatory Control

The essence of this type of continuous control is to regulate or keep a particular process performance at a certain level or within a specified tolerance limit. The limitations of regulation control are that the control will not take any action unless there is an error or disturbance and compensation action is only possible after the disturbance has affected the process.

Feed Forward Control

Feedforward control can anticipate, sense, and compensate for a disturbance before it offsets the balance of a system. This type of control is proactive in nature and would take action before there is a disturbance or error. The advantage is that it can prevent or detect the presence of any potential disturbance, which could affect the process, and the feedback control prevents such a disturbance from affecting the process. This control can take corrective action once a potential disturbance is detected by adjusting the process parameter to compensate for the effect such disturbance has on the process. However, the control cannot completely compensate for the error or disturbances in the system.

Steady-State Optimization

This type of control operates as an open loop and can be used for process optimization to ensure optimum production rate and production of high-quality products. This control can enhance the determination of the process variable and the system parameter value for optimization. The merit of this control is that the control can adjust the process parameters to drive the process towards the optimal state, hence, it can be useful for process optimization. However, the presence of errors or disturbances can affect this type of control.

Adaptive Control

This is an open-loop, self-controlling model of continuous control, which combines feedback and optimal controls. Adaptive control can measure certain process variables.

Depending on the objective of the control, adaptive control can perform the following functions:

1. Identification function: Since the process is dynamic, continuous control can assist in measuring the current value of a particular process performance.

2. Decision function: Once the current value is determined, continuous control can assist in making decisions on the course of action to keep the system performance at the desired level.

3. Implementation function: To execute the decision taken for the improvement of the process variable usually with the aid of an actuator.

A successful control defines the desired performance or system's behaviour, generates and applies actions where necessary, and selects actions to make modifications.

Examples of industrial control systems are the process control system and speed control system amongst others.

PROCESS CONTROL SYSTEM

This is the technology of controlling a series of activities to transform a raw material into a desired end product [3]. This aim is to regulate the process variable for the optimum performance of a system. The process control systems adjust to compensate for external disturbances that can affect the system's behaviour. Examples of controls that belong to this class include; temperature, pressure, liquid level, and flow rates control systems.

Temperature Control System

The temperature due to the passage of heat in and out of the system can be detected *via* a temperature probe and adjusted to the threshold value with the aid of a thermostat. Thermostatically controlled devices include the following; boilers, refrigerators, air conditioners, pressing iron, *etc.*

There are four commonly used temperature sensor types:

Negative Temperature Coefficient (NTC) Thermistor

A thermistor is a heat-sensitive resistor that produces a large, predictable, and precise change in resistance which correlates to variations in temperature. An NTC thermistor gives a very high resistance at low temperatures and as the temperature increases, the resistance also drops quickly. This is due to the fact that for the NTC thermistor, a large change in resistance per temperature change is reflected very fast and with a high degree of accuracy (say 0.05 to 1.5 °C), hence, the output of an NTC thermistor requires linearization due to its exponential nature. The effective operating range is often between -50 to 250 °C for glass-encapsulated thermistors or 150°C for standard.

Resistance Temperature Detector (RTD)

An RTD is also known as a resistance thermometer and is used to measure temperature by correlating the resistance of the RTD element with temperature. It consists of a film, a wire wrapped around a ceramic or glass core for greater accuracy. RTDs made of platinum are very accurate, however for cost considerations, RTDs lower in cost and accuracy can be made from nickel or copper. RTDs made of platinum offer a fairly linear output that is highly accurate in the range of 0.1 to 1 °C across -200 to 600 °C temperatures.

Thermocouple

A thermocouple consists of two wires of different metals connected at two points. The difference in voltage between these two points reflects the corresponding changes in temperature. Thermocouples are nonlinear hence it requires conversion when used for temperature control and compensation. This is often accomplished using a lookup table. The accuracy of a thermocouple is often low ranging from 0.5 °C to 5 °C but has the merit of being operated over a wide range of temperatures from -200 °C to 1750 °C.

Semiconductor-Based Sensors

Semiconductor-based temperature sensors are often placed on integrated circuits (ICs). These sensors consists of two identical diodes with temperature-sensitive voltage and current characteristics which is used for monitoring temperature changes. This type of temperature sensor offers linear responses, however its demerits often include poor accuracy, slow responsiveness (5 s to 60 s) and it operate over a narrow range of temperature (-70 °C to 150 °C).

Pressure Control System

The components of a pressure control and measurement are shown in Fig. (**1**). The flow capacity is controlled by the control valve while the desired pressure is set through the ethernet. The control valves help to keep the system pressure below the desired upper limit thereby maintaining set pressure in a part of the circuit.

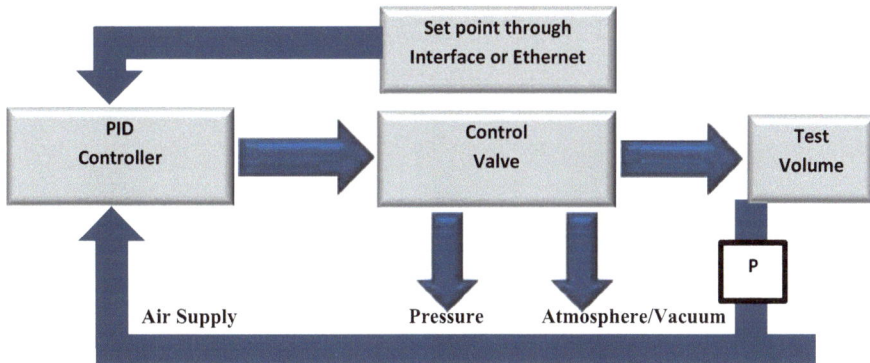

Fig. (1). Pressure control system.

Liquid Level Control System

Flow switches are often employed to monitor and control water liquid level while the oil level sensors using magnetic reed switches are used to measure oil levels and automatically turn on or off oil pumps. Just like the principle of operation of a float switch, the reed switch moves up and down the stem to open or break circuits (turn on or off oil pumps) according to oil levels rising and falling. Once the oil level in the tank assumes its lowest predetermined point (closed position), a circuit will be created by the reed switch, which will send an automatic signal to the pump to resume the pumping action until it fills up the oil tank to the predetermined level. The magnetic reed switch will then open the circuit again (open position) once the oil level reaches the maximum fill capacity.

Flow Rates Control System

The flow rate control system is used to control or adjust the amount of fluid flow in a system. It comprises a flow meter, which measures the amount or quantity of flow. It has an orifice as a small passage for flow. This is connected to a flow transducer that measures the velocity of the flow. An adjustable control valve connected to a controller is used to regulate the rate of flow.

Speed Control System

This responds to input commands to regulate the physical motion as well as the system's position. For example, the rotary arm or the robotic arm that performs welding operations in an industry, the Computer Numerical Control in the machine (CNC) amongst others.

TYPES OF CONTROL SYSTEMS

The following are types of control systems [4 - 6].

1. Open Loop Control System

2. Closed Loop Control System

3. Open and closed Loop Control System

Open Loop Control System

This type of control system manipulates without getting feedback from the system (*i.e.* they are non-feedback control systems). Hence, the control action is independent of the output of the system whose behaviour is being manipulated as the controller cannot compensate for changes in the system. In other words, for an open loop control system, the output has no effect upon the input variable in the control process. They are mostly employed when the outcome of the process is certain since there is no automatic correction when there is variation in the system's behaviour. Open loop controls are non-feedback systems, usually managed by human intervention where the operator observes the key variables such as system power, pressure, or level and then makes manual adjustments to the controls in order to achieve the desired result. Examples of operations involving open loop control systems are: opening a door, shutting off a tap or pump, water reservoir for irrigation purposes, automatic washing machine, some traffic robots, a central heating boiler such that the heat is applied for a constant time irrespective of the output temperature having the control action as a simple ON/OFF switch *etc.* An open loop control system can be represented as shown in Fig. (**2**).

Fig. (2). Block diagram of the open loop control system.

The following are advantages of the open loop control system:

1. They are cheaper and simple to install and operate.

2. Open loop control systems require less maintenance operations.

3. They are often simple to build and calibrate.

However, their disadvantages lie in the fact that;

1. They are not accurate and reliable since they manipulate systems' performance without feedback. Without the feedback, there is no assurance that the desired process or output variables will meet the target.

2. They are often characterized by poor precision and accuracy.

3. The optimization of process or output variables is not always possible.

Closed Loop Control System

For this type of system, the control system gets automatic feedback as the behavior of the system is manipulated. There is a flow of information around a repetitive feedback loop until the processor output variables are kept within the desired range. A sensor monitors the system's condition and feeds the data obtained to a controller, which adjusts the output device as necessary to maintain the desired system output. The control loop comprises a feedback loop where the state of the system is fed back into the controller and compared with the reference point to provide an error signal to the controller, which makes the necessary changes in order to offset or compensate for such error to keep the process conditions within the desired or predetermined points. Therefore, the feedback loop is an essential component of the closed-loop controller which ensures that part of the output is fed back into the system to form part of the input or system's excitation and the controller exerts a control action to give a process output that is within the range of the "reference input" or "set point". Hence, because of this feedback loop, closed loop controllers are also referred to as feedback controllers. The precision and accuracy of control in terms of the closeness of the output to the set point are a function of the feedback path. The output of the system affects the input as there is automatic correction by the control system until the error signal is negligible. The error signal of a closed-loop control system is the difference between the input signal and the feedback signal (which may be the output signal itself or a function of the output signal). The error signal is what is fed into the controller which affects the control and keeps the subsequent output of the system within the desired range of value. It is chiefly employed when the output of a process is subjected to variation over some time interval. Examples of operations involving the closed control loop are; speed control in automobiles, traffic robots, temperature, pressure flow, and fluid level control in a reactor. The illustration of the block diagram of the closed-loop control system is shown in Fig. (3) [7].

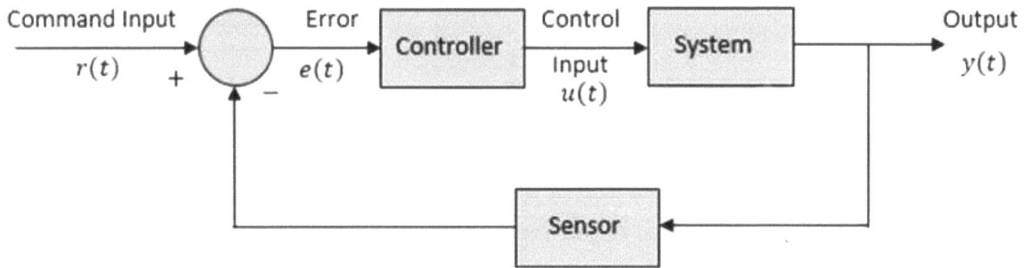

Fig. (3). Block diagram of the closed-loop control system.

For instance, in an automobile, the speed of the car is the process variable that needs to be manipulated. The speedometer acts as the sensor, the cruise control is the controller, and the accelerator is the actuator that manipulates the speed. The desired speed (threshold speed) is set on the cruise control which maintains the speed of the car over time. The speedometer senses the speed and the value is compared with the threshold value. If the value is higher than the threshold, the controller disengages the accelerators until the set point is met. On the other hand, if the value is lower than the threshold, the accelerator is engaged more until the set point is met.

Another example involves the electric iron used for removing wrinkles or creases from a cloth. The control of modern ones is often incorporated with a feedback loop. It has a temperature sensor that measures and monitors the dryness of the clothes and compares it with the input reference. The error signal is amplified by the controller, and the output from the controller automatically adjusts the heating system to reduce any error. The adjustment of the error prevents overheating of the electric iron which may burn the clothes or insufficient heat transfer which may prevent the removal of wrinkles from the clothes. The automatic adjustment from the controller causes the heating element to trip off when the threshold value is exceeded and to turn on when the temperature falls below the threshold.

The following are the advantages of the closed-loop control system;

1. They are reliable, fast, and can as well enhance repeatable performance in operation.

2. They operate with a high degree of precision and accuracy.

3. Process optimization is often possible with the closed-loop system.

4. Reduction of the system's sensitivity to external disturbances *via* adequate error compensation.

They pave the way for automatic error adjustment.

6. Highly efficient in ensuring the stability and robustness of a system against external disturbances.

The following are the disadvantages of the closed-loop control system;

1. They are more expensive than the open loop control system.

2. The fact that it receives feedback from the system being manipulated makes the control system and installation complex with the addition of one or more feedback loops.

3. They often require maintenance activities more than the open-loop control system.

4. It takes more time to execute control when compared to the open loop system. This is due to the fact that the process of variable measurement, conversion to signal, error generation, transmission and comparison takes time.

5. When the gain of the controller is too sensitive to changes from the input signals, it can become unstable producing undesirable oscillations.

Closed loop systems are generally divided into two namely;

1. Continuous closed loop system

2. Discrete closed loop system

The control loop of the continuous process is more complex than the discrete. The discrete system instantaneously measures the control variable when there is physical observation of deviation from the set points and executes control. For instance, a driver turns on the light of a car when the ambient light seems too dark to drive. On the other hand, the continuous closed-loop system uses the feedback loop to keep the output variable within the desired range. The value of the error generated and sent to the controller is compared with the threshold before the controller decides on the actions needed to maintain the output variable within the desired range. The process is repetitive until the desired range of operating or output variables is met. The use of cruise control in an automobile to maintain the speed of a car is an example of a continuous closed loop system.

The following are some of the variables that can be controlled in a system; temperature, flow level, and pressure amongst others.

Closed Loop Temperature Control in a Reactor

The temperature of a reactor can be best controlled using a closed loop system. For instance, if the heating element is operated by a 220 V power supply, a thermocouple can be connected to the heating element to measure the temperature. A thermocouple amplifier can also be connected to the output of the thermocouple to amplify the voltage coming out of the thermocouple. Therefore, the amplified voltage from the amplifier is then passed through a microcontroller where the value is converted into digital form using Analog to Digital Converter (ADC) on the microcontroller. An Analog to Digital Converter (ADC) converts continuous-time analog signals to discrete-time signals in order to obtain a representative set of numbers that can be used by a digital computer. On the controller, the threshold temperature is set to the desired operating temperature of the reactor and the measured voltage is compared to the acceptable voltage range. If the measured temperature of the heating element is lower than the threshold, the micro controller turns ON the contactor which in turn activates the heater. On the other hand, if the measured temperature is greater than or equal to the desired threshold, the microcontroller turns OFF the contactor which in turn, puts OFF the heating element. The microcontroller can be used in conjunction with an internal timer to control a buzzer to essentially raise an alarm through a transistor circuit configured as a switch to alert the personnel whenever an urgent action is needed [8].

Closed Loop Temperature Control in a Motor

In order to achieve the control of a motor in a closed loop, a speed measuring transducer, such as a tachometer will be connected to the shaft of the DC motor for speed measurement. The detected speed signal will be sent to the amplifier. The input speed, θ_i which is amplified by the controller to drive the DC motor is pre-set and the speed N which represents the output, θ_o of the system, as well as the tachometer T would be the closed-loop back to the controller. The difference between the input voltage and the feedback voltage gives the error signal as shown in Fig. (**4**).

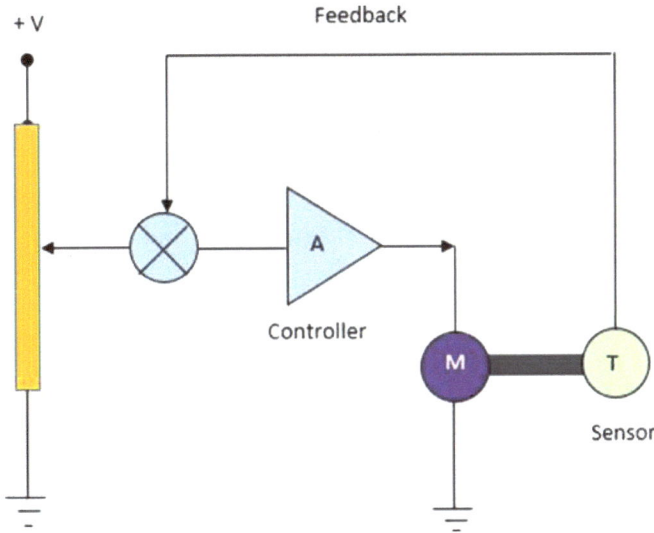

Fig. (4). Closed loop temperature control in a motor.

Any external disturbances to the closed-loop motor control system such as the motors load variation would create a difference in the actual motor speed and the input set point.

This difference would produce an error signal which the controller would automatically adjust in order to control the motors speed. Then the controller works to minimize the error signal, with zero error indicating actual speed which equals the set point.

This simple closed-loop motor controller can be represented as a block diagram as shown in Fig. (**5**).

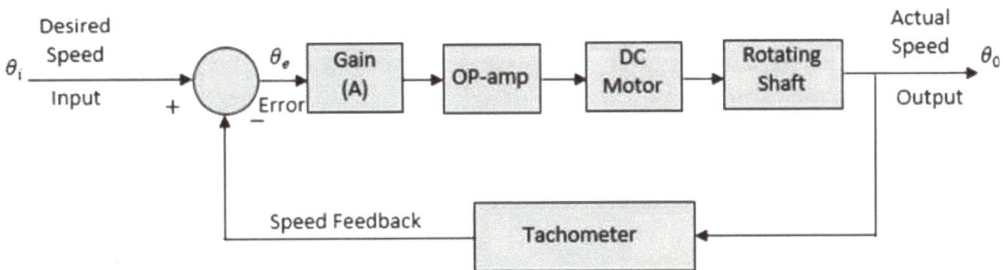

Fig. (5). Block diagram of closed-loop motor controller.

Open and Closed Loop Control System

The feedback system is a system which maintains a preset condition of the system in relation to other variables by comparing the magnitude of the variables and obtaining the error difference as a basis for adjustment and control [9, 10]. The system is continuous and involves the use of sensors for measurement and subsequent adjustments in order to maintain the ideal conditions, for example, the control systems of aircraft, advanced manufacturing industries, some automobiles, *etc.* When there is a need to combine speed with accuracy, the open and closed-loop control system is most suitable. It uses a feed-forward control mechanism that simulates the system in the open loop. The error generated as a result of deviation from set points is adjusted. The feed-forward control also senses impending disturbances, which can offset the balance of the set points. These disturbances are therefore controlled or eliminated before any effect on the system. Examples of systems using the open-closed loop systems are automobile and robotic vehicles with an obstacle alarm and avoidance sensors.

Feedback is an important component of the control system due to the following reasons [9, 10].

1. It reduces error between the actual and the desired values. When the error generated exceeds the permissible limits, the controller manipulates the input signals in such a way as to minimize the error. On the other hand, the system can be trouble-shot to examine the possible causes of the error.

2. Reduces the effect of variation of system parameters and output, therefore, enhancing the stability of the system. Frequent undesired changes in process parameters and systems' output are minimized with the help of a feedback system which monitors and reports systems' performance with respect to the reference signal.

3. Reduces the effect of noise and other disturbances: Noise and other disturbances that could alter the course of a system are reported by the feedback system and checked by the controller.

4. It increases the overall system performance in terms of precision, accuracy, as well as the overall system gain and bandwidth.

THE SUMMING POINT OF A CLOSED LOOP SYSTEM

For a closed-loop control system to effect control, the error between the actual output and the desired output must first be determined through an error signal which is fed back into the system. This is achieved with the aid of a summing point, called the comparator or a comparison element, incorporated between the feedback loop and the systems' input. At the summing points, the set points are compared to the actual value to produce a positive or negative error signal which the controller responds to in real time to effect control (Fig. **6**).

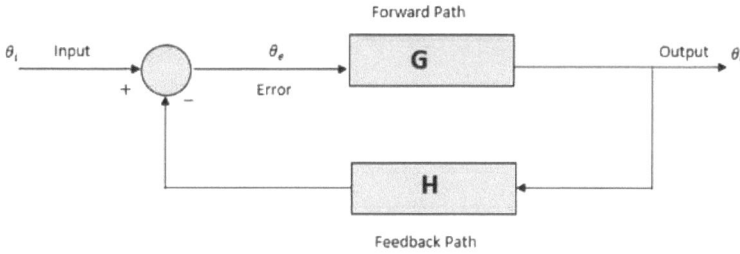

Fig. (6). The summing of closed loop system.

The error (e) is expressed as Eq. (1).

$$e = V_s - V_a \qquad \qquad (1)$$

Where:

V_s is the set value and V_a is the actual values.

Where: The block G represents the open-loop gains of the controller in the forward path, and block H represents the gain of the sensor in the feedback path.

The symbol used to represent a summing point in closed-loop systems block-diagram is a circle with two crossed lines as shown in Fig. (**6**). The summing point can either add signals together indicated with a plus (+) symbol as in the case of a "summer" which is employed for positive feedback, or it can be a subtracted signal indicated with a minus (−) symbol as in the case of a "comparator" which is used for negative feedback.

THE TRANSFER FUNCTION OF A CLOSED LOOP CONTROL SYSTEM

The relationship between the systems' input and its output is usually represented by a mathematical expression called the transfer function. This is used for describing the behaviour of such a system. The gain of the system is the ratio of the output to the input. Therefore for a closed loop system, the output of the system is a product of its transfer function and the input.

In order to determine the transfer function of the closed-loop system illustrated above, there is a need to first determine the output signal θ_o in terms of the input signal θ_i. This can be done as follows;

The output (θ_o) from the system is expressed as Eq. (2):

$$\theta_0 = G \times \theta_i \tag{2}$$

The error signal (e) is also the input fed as gain in block G and (θ_i) is the input signal.

The output from the summing point is expressed as Eq. (3).

$$e = \theta_i - H \times \theta_0 \tag{3}$$

If the gain of the sensor (H) = 1 (unity feedback) then:

The output from the summing point is expressed as Eq. (4).

$$e = \theta_i - \theta_0 \tag{4}$$

Eliminating the error term, then the output is expressed as Eq. (5).

$$\theta_0 = G \times (\theta_i - H \times \theta_0) \tag{5}$$

Hence, expressing the output as a product of the gain and input, Eqs. (5) and (6) apply;

$$G \times (\theta) = \theta_0 + G \times H \times \theta_0)$$

$$\tag{6}$$

$$\frac{Output}{Input} = \frac{\theta_0}{\theta_1} = \frac{G}{1+GH}$$

Eq. (6) is the transfer function of a closed-loop system with a negative feedback as indicated by the plus (+) sign in the denominator. For a closed loop system with a positive feedback with minus (−) sign at the denominator, Eq. (7) holds thus;

$$\frac{Output}{Input} = \frac{\theta_0}{\theta_1} = \frac{G}{1-GH} \qquad (7)$$

The feedback is unity when H = 1 and when G is very large, the transfer function approaches unity.

A decrease in the magnitude of gain G produces a corresponding slow decrease in the ratio of the output to input thus making the system fairly insensitive to external disturbances and internal variations.

ELEMENTS OF THE CONTROL LOOP SYSTEM

The key elements of the control loop are shown in Fig. (7).

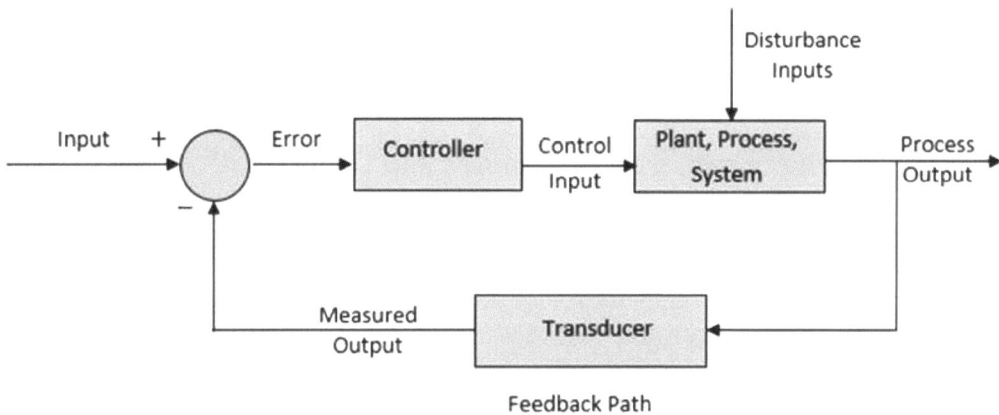

Fig. (7). Elements of a control loop.

The following are the elements of the control system;

1. System: This generally refers to a physical process which can be modelled, observed or measured.

2. State: This describes the task performed by the system. The state of a system is the work performed by the system.

3. Dynamics: This describes how the behavior or performance of a system changes with time.

4. Reference Signal: This describes the projected performance of the system. In other words, the expected output from the system.

5. Control Input Signal: This is the input signal being manipulated by the control system.

6. Output Signal: This is the variable signal that has been manipulated.

7. Actuator: It is a control element which gives executes instructions from the controller by giving the output of an action that is meant to change the control variable. The actuator also converts systems output into physical movements. Examples of actuators are: valves, servo motors, pumps, relays, speed drives, bakes, relays, solenoids *etc.*

The sensor measuring device measures the condition of the process as well as the output of the system. It provides input from the process and from the external environment. It can also acquire information on the progression of process variables and converts the measured or acquired information into electrical or optical signals.

1. Transducer: It converts non-electrical signal into an electrical value.

2. Transmitter: It converts the measurement into an electronic or optical signal and sends the signal.

3. Comparator: This is a feedback element which measures the differences between the reference signal and the sensor's output. The error generated is thereafter sent to the controller.

4. Controller: The threshold value also known as the desired value or set point is set on the controller, the controller decides whether or not the process is acceptable.

5. Variable: This is the quantity or process parameter being detected or measured. They are the quantities that are expected to be kept at the desired range below the threshold so that process and system will perform as expected with permissible deviations. However deviations from set points tend to exist when external disturbances influence the variables causing a shift from the desired or set points. Examples of such process parameters are viscosity, flow, temperature, load, density, liquid level, pressure, voltage, inductance, resistance, capacitance, vibration, weight, current, frequency *etc.*

6. Error Signal: This an electrical signal representing the difference between the desired and true output signals.

There are various types of closed loop control systems being used in control applications. They are:

1. Single Input Single Output (SISO): Both the signals to be manipulated as input and the emerging manipulated signal as output are considered one.

2. Multiple Input Single Output (MISO): More than one signal is to be manipulated as input with only one emerging manipulated signal as output.

3. Multiple Input Multiple Output (MIMO): Both the signal coming in to be manipulated and the emerging manipulated signal as output are more than one.

4. Single Input Multiple output (SIMO): Only one signal coming in to be manipulated while many manipulated signals will emerge as output.

TYPES OF VARIABLES

Usually there are three types of variables in any control system. They are;

1. Controlled variables

2. Manipulated variables

3. Disturbance variables

Controlled Variable

These are otherwise referred to as the output variable. These variables determine the performance of the system.

Manipulated Variable

These are the variable being adjusted dynamically for optimal performance of the system.

Disturbance Variables

These are external or load variables, if not controlled, they can cause the controlled and manipulated variables to deviate from their set points. Disturbance variables may be due to environmental factors that influence the systems' behaviour. For example wing gradient.

CLASSES OF CONTROL

Control systems generally fall into three main categories depending on the method of actuation namely;

1. Manual control systems

2. Semi-Automatic control systems

3. Automatic control systems

Manual Control Systems

The actuation mechanism relies solely on human efforts for execution.

Semi-Automatic Control Systems

The actuation mechanism relies partially on both human effort and programs for control. Some systems require human effort for initiation while they can complete the task of executing the control on their own. Sometimes, operators' intervention is needed to keep control variables within the desired range when failure occurs from the one or more feedback mechanisms such as the sensors, actuators, transmitters, *etc.*

Automatic Control Systems

These are systems that do not rely on any level of human intervention for their control from the initiation to the execution. Examples are the space vehicle, missile guidance, robots *etc.* Some automatic control systems have a manual mode which can be activated in any case of failure. The feedback loop can be broken to switch to the manual mode.

TYPES OF SYSTEM PROCESSES

The kind of control system employed however largely depends on the type of process used by the system to be controlled. For instance, systems' processes can be executed batch-wisely, continuously or individually.

Batch Processing

In a batch processing system, all the reactants are combined in a single process vessel to produce one batch of product at a time. The reaction mixture is fed batch-wisely and sufficient reaction time is allowed for conversion into a finished product under desired reaction conditions (temperature, pressure and stir rates). Once the predetermined reaction time is complete, the contents of the system are removed and sent for subsequent processing for another batch to be fed in. The batch process has several positive features including good mixing characteristics, relative ease of handling raw materials and operational flexibility. However, batch reactors are generally not used in the production of large volumes of products since it is most efficient to operate the subsequent purification and packaging steps in a continuous mode.

Continuous Processing

In continuous processing, the process runs continuously with raw material fed as input from one end and the finished product from the other end. There is a steady flow of raw materials into the system and products out of the system. Once a continuous flow system reaches steady-state operation, the product composition leaving the system becomes constant. Continuous systems are generally large commercial plants which process the raw materials and produce the finished product in a continuous flow. A continuous plant leads to better heat economization, better product purity minimal operator interference in adjusting plant parameters, and lower capital cost per unit of finished product. The continuous plant also requires lesser residence time for the completion of the process compared to the batch processing plants.

Individual Processing

For this type of processing, a series of operation procedures produce the useful product or individual product.

System's Error

These are uncertainties that accompany the measurement of system input or output variables. It is an electrical signal representing the difference between the desired and the true output signals.

Basically, systems' errors are of two types;

1. Error due to precision: this is related to the random error distribution that is associated with a particular operation.

2. Error due to accuracy: this is related to the existence of a systematic error *e.g.* incomplete calibration, *etc.*

REPRESENTATION OF THE CONTROL SYSTEMS

Control systems can be represented by any of the following models;

1. Differential equations;

2. Laplace equations;

3. Block diagrams;

4. Signal flow graphs.

PROPORTIONAL–INTEGRAL–DERIVATIVE (PID) CONTROLLER

As the name implies, the PID controller consists of three basic terms; proportional, integral and derivative which are varied to get an optimal response. The proportional, integral and derivative controllers are combined as one in one control feedback loop (controller) often used in industrial control systems. In the loop, the sensor is used to measure the process parameter, then the desired actuator output is calculated by determining the proportional, integral, and derivative responses and summing those three components to compute the output. The controller continuously calculates the value of the error signal which is the difference between the preset desired value and the measured process variable and applies an appropriate correction based on the proportional, integral, and derivative terms, respectively. At any given moment, the error is used by the control system algorithm (compensator), to determine the desired actuator output to drive the system. For instance, in a boiler system where a thermocouple is used as a temperature sensor, if the desired set point of the process temperature is 80 °C and the measured temperature reads 50 °C, the heater through the microcontroller will be activated to raise the temperature until it is 80 °C. On the other hand, if the measured temperature is greater than the desired set point (less than 80 °C), the microcontroller deactivates the heater.

This is called a closed loop control system because the process of reading the signal from the sensor to provide constant feedback and calculating the desired actuator output is repeated continuously.

The PID controller finds application in the precise control of a system where a definite output control is desired. Therefore, continuous monitoring of such a system is required. Examples of such systems include; the fuel tank used in a boiler-based power plant amongst others. This type of control is referred to as modulating control. Systems with feedback mechanisms and Proportional-Integral-Derivative (PID) controllers are often employed for the control of industrial systems.

Fig. (**8**) shows the block diagram of a PID controller in a feedback loop.

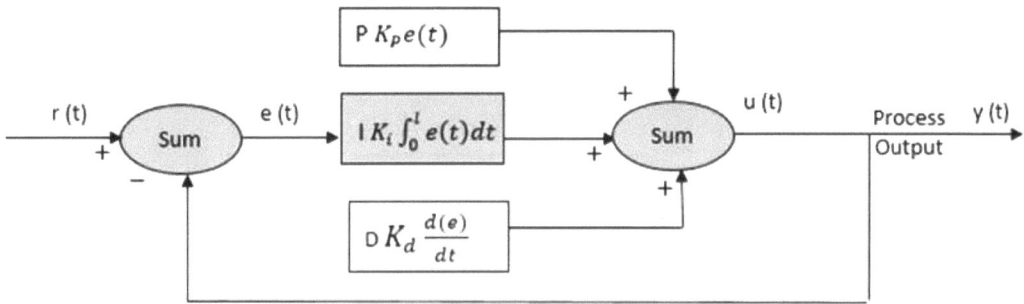

Fig. (8). A PID controller in a feedback loop.

Where;

r(t) is the desired process value or set point, et is the error signal, ut is the output value expressed as a function of the input value and y(t) is the measured process value.

For a PID controller, the output can be expressed in terms of input as Eq. (8).

$$u(t) = K_p e(t) + K_1 \int_0^t e(\tau)d\tau + K_d \frac{de(t)}{dt} \qquad (8)$$

Therefore, the transfer function of PID controller can be written as Eq. (9).

$$y(t) = K\left(1 + \frac{1}{\tau s} + tS\right) \qquad (9)$$

Where:

K_p is the proportional gain; τ is the integral time and t is the derivative time, K1 is the proportional gain, τ is the integral time and is the derivative time.

The output of the proportional controller is always proportional to the error value with a gain value of K_p and the proportional response can be adjusted by multiplying the error by the proportional gain constant K_p. For an integral controller, the manipulation equals the integral of the error over a time interval, multiplied by a gain K_1 while the derivative controller uses the derivative of the error instead of the integral. The proportional component depends on the error term (*i.e.* difference between the set point and the process variable). For instance, if the error term has a magnitude of 5, a proportional gain of 2 would produce a proportional response of 10. Generally, an increase in the proportional gain increases the speed of the control system response and vice versa. However, large values of the proportional gain force the process variable to oscillate. A further increase in the value of the proportional term may bring about large oscillations which will offset the balance of the system. The integral component computes the error term over time. This implies that a small error term will cause a slow increase in the integral component. The integral response will continually increase over time until the error becomes zero. The effect of the increase in the integral term is to drive the steady-state error to zero. The steady-state error is the final difference between the process variable and set point. The derivative component causes the output to decrease with increasing process variable having the derivative response directly proportional to the rate of change of the process variable. An increase in the derivative time (T_d) will bring about an increase in the speed of the overall control system response as the control system will react more strongly to changes in the error term. Most practical control systems often use small derivative time (T_d), because of the high sensitivity of the derivative response to noise in the process variable signal. The generation of noise by the feedback signal or slow rate of the control loop rate will make the control system unstable. For effective control, there is need to set the optimal gains for P, I and D so as to get an ideal response from a control system in a process called "tuning". However, the process of setting the gains of a PID controller can be an iterative process of trial and error. In this method, the "I" and "D" terms are first set to zero while the proportional gain is increased until the output of the loop begins to oscillates. An increase in the proportional gain increases the speed of the system while a further increase may cause the system to go unstable. Hence the proportional term can be set to obtain a desired fast response while the integral term is increased to stop its oscillating effects. The integral term often reduces the steady state error but increases the magnitude of overshoot. However, some amount of overshoot is always necessary for a fast system for a quick response to changes. Once the proportional and integral terms have been set to get the desired fast control system with minimal steady-state error, the derivative term can be increased until the loop is acceptably quick to its set point. Hence, an increase in the derivative term brings about a significant reduction in the overshoot and with higher stability. The only demerit is the fact that the system may be highly sensitive to noise. Often times, for effective control, there is a need to trade off one characteristic of the control system for another in order to meet the desired requirements.

The Ziegler-Nichols method is another popular method often employed for tuning a PID controller. In this method, the integral and the derivative terms are set to zero while the proportional term is increased until the loop starts to oscillate. The P, I and D terms are then adjusted according to Table **1**.

Table 1. Ziegler-Nichols tuning method.

Control	P	Ti	Td
P	0.5Kc	-	-
PI	0.45Kc	Pc/1.2	-
PID	0.60Kc	0.50Kc	Pc/8

The control system's performance is measured by applying the set point, and measuring the response of the process conditions in the form of a wave characteristics. The rise time is the amount of time it takes the system to go from 10% to 90% of the steady-state, or final, value while the percent overshoot is the amount that the process variable exceeds the final value, expressed as a percentage of the final value. Also, the settling time is the time required for the process variable to settle and regain its balance within a certain percentage (commonly 5%) of the final value after encountering a disturbance. The time between the period of changes in the process variable and when the change can be observed is referred to as dead time. For instance, some sensors such as the temperature or flow sensors may not measure a change in the process conditions immediately. This can be as a result of a slow response of the system's actuator to the control instructions from the controller. Some systems that change quickly or have complex behaviour require faster control loop rates. PID controllers can either be configured in series or in parallel. For series configuration, the output of one control depends on the output of the other while in parallel configuration, the output of one controller is independent of the output of the other controller.

The applications of PID controllers in the manufacturing industry are as follows;

1. In automobiles, PID control is used in the control of the active suspension system as well as the automatic car steering and it provides efficient control when integrated with Fuzzy Logic or other forms of adaptive controls.

2. In movement detection system of the modern seismometer.

3. In water/oil level monitoring in tanks.

4. Head positioning of a disk drive.

5. Automated inspection and quality control.

6. Manufacturing process control: CNC machine tools.

7. Chemical process control: flow control, temperature control.

8. Automatic control of material handling equipment.

9. Automatic packaging and dispatch.

10. To ensure safety during manufacturing operations.

THE CONTROL THEORY

The control theory is a branch of engineering that deals with the design and analysis of control systems as well as the behaviour of dynamic systems; their inputs and modifications of the inputs to obtain a desired output. The control theory is basically divided into two types: the linear control theory and the nonlinear control theory. The linear control theory assumes a linear model and the system follows the principles of superposition. The principles of superposition state that for a linear system, the output of a sum of inputs is equal to the sum of the respective outputs. In other words, the response of a linear system to a sum of signals is the sum of the responses of each individual response input signal [11 - 13].

Table **2** presents the differences between a linear and nonlinear control systems.

Table 2. Differences between a linear and nonlinear control systems.

Linear control systems	Nonlinear control systems
Exist only in theory.	Exist in real world systems. All control systems are nonlinear in nature.
Assume a linear model and the system follows the principles of superposition.	They do not assume a linear model, thus, they are linearized through approximation and thereafter a linear technique is employed for solving it.
Governed by the principles of linear differential equation.	Governed by the principles of non-linear differential equation.
Solved *via* Laplace transformation, Fourier or Z transform, bode plot, root locus, Nyquist stability criterion.	Solved *via* numerical methods, simulation. *etc.*
Deals with linear time invariant systems	Deals with systems that are non-linear, time variant systems or both.

Linearization

Linearization involves the linear approximation of a nonlinear system that is valid in a small region around an operating point.

For dynamic systems, the continuous time non-linear differential equation can be expressed as Eqs. (10a and b).

$$(\dot{x})t = f(x(t), u(t), t) \tag{10a}$$

$$y(t) = g(x(t), u(t), t), \tag{10b}$$

From Eqs. (10a and b), x(t) denotes the system states, while u(t) denotes the input into the system and y(t) denotes the output of the system.

The linearized model of the system expressed in Eq. (10) is given in Eq. (11).

$$t = t_o, x(t_o) = x_o, u(t_o) = u_o \ and \ y(t_o) = g(x_o, u_o, t_o) = y_o \tag{11}$$

The variables of the linearized model are defined by Eqs. (12-14).

$$\delta \dot{x}(t) = x(t) - x_o \tag{12}$$

$$\delta u(t) = u(t) - u_o \tag{13}$$

$$\delta y(t) = y(t) - y_o \tag{14}$$

The linearized model in terms of Eqs. (12 -14) are given in Eqs. (15) and (16).

$$\delta \dot{x}(t) = A \delta x(t) + B \delta u(t) \tag{15}$$

$$\delta y(t) = C \delta x(t) + D \delta u(t) \tag{16}$$

Control Feedback Linearization

There is no general method of the design of a non-linear controller and one of such methods is the feedback linearization.

Recall Eq. (10a),

$$(\dot{x})t = f(x(t), u(t), t) \tag{10a}$$

The input state transformation is expressed as Eq. (17).

$$u = g(x, m) \tag{17}$$

The non-linear system is changed into an equivalent linear time-invariant dynamic system as expressed by Eq. (18).

$$\dot{x} = Ax + bm \tag{18a}$$

$$m = -Kx \tag{18b}$$

$$\dot{x}_1 = x_2 \tag{18c}$$

$$\dot{x}_2 = x_3$$

$$\dot{x}_{n-1} = x_n$$

$$\dot{x}_n = f(x) + b(x)u \tag{19}$$

$$u = \frac{1}{b(x)}(m - f(x))$$

$$\tag{20}$$

$$b(x) \neq 0$$

Where: u is the scalar control input, f and b are non-linear function of the state.

To cancel the non-linearity and impose the desired linear dynamics, Eq. (20) holds thus;

For $f(x) \, \epsilon X$ - *State space*

The linear model is obtained thus:

$$\dot{x}_1 = x_2$$

$$\dot{x}_2 = x_3$$

$$\dot{x}_{n-1} = x_n$$

$$x_n = m$$

The illustration of the input state linearization is shown in Fig. (**9**).

$$\dot{x} = Ax + bm$$

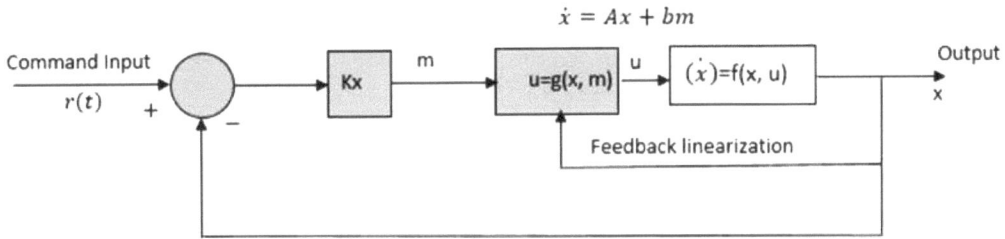

Fig. (9). Input state linearization.

CONCLUSION

A control system is a system which manages, instructs, regulates and directs the behaviour of a simple device or a more complex system with the aid of the control loops (which can be open, closed or combination of both). A control system can range from a simple domestic control such as a thermostat for controlling the boiler or the electric iron to a complex industrial controller for controlling systems, processes or machines. Control actions can be discrete or continuous while the levels of control can range from manual control to semi-automatic to fully automatic controls depending on the level of human involvement in the execution of control actions.

REFERENCES

[1]   N.E. Battikha, *The Condensed Handbook of Measurement and Control* 2nd ed.. ISA, 2004, pp. 1-8.

[2]   S. Parthasarathy, "Is there a relation between sequential control and continuous control?", *Math. Comput. Model.,* vol. 13, no. 1, pp. 55-59, 1990.
[http://dx.doi.org/10.1016/0895-7177(90)90114-3]

[3]   W.C. Dunn, "Introduction to instrumentation, sensor and process control", In: *Industrial Process Automation Systems,* B.R. Mehta, Y.J. Reddy, Eds., Artech House Inc.: Norwood, 2015, pp. 1-348.

[4]   K. Warwick, *An Introduction to Control Systems.* World Scientific: Singapore, 1996.
[http://dx.doi.org/10.1142/2175]

[5]   D.R. Coughanowr, *Process Systems Analysis and Control.* McGraw-Hill, Inc.: London, 1991.

[6]   W. Bolton, *Instrumentation and Control Systems.* Elsevier Science & Technology Books, 2004.

[7]   M. Laskawski, and M. Wcislik, "Sampling rate impact on the tuning of pid controller parameters", *Int. J. Electron. Telecommun.,* vol. 62, no. 1, pp. 43-48, 2016.
[http://dx.doi.org/10.1515/eletel-2016-0005]

[8]   I.A. Daniyan, K. Mpofu, L. Daniyan, and A.O. Adeodu, "A.Development and automation of a multi-feedstock plant for biodiesel production", *International Conference on Electrical, Computer and Energy Technologies (ICECET),* 2021pp. 1-9 Cape Town
[http://dx.doi.org/10.1109/ICECET52533.2021.9698675]

[9]   J. Love, *Process Automation Handbook:A Guide to Theory and Practice.* Springer-Verlag London Limited: London, 2007.

[10]   D.E. Seborg, T.F. Edgar, and D.A. Mellichamp, *Process Dynamics and Control.* John Willey & Sons, Inc.: USA, 2004.

[11]   M. Vidyasagar, *Nonlinear Systems Analysis.* 2nd ed. Prentice Hall: Englewood Cliffs, 1993.

[12]   A. Isidori, *Nonlinear Control Systems.* 3rd ed.. Springer: Berlin, 1995.
         [http://dx.doi.org/10.1007/978-1-84628-615-5]

[13]   H.K. Khalil, *Nonlinear Systems.* 3rd ed.. Prentice Hall: Upper Saddle River, 2022.

CHAPTER 6

Computer Control Devices in Automation

Ilesanmi Afolabi Daniyan[1,*], **Lanre Daniyan**[2], **Adefemi Adeodu**[3] and **Felix Ale**[4]

[1] *Department of Industrial Engineering, Tshwane University of Technology, Pretoria 0001, South Africa*

[2] *Department of Instrumentation, Centre for Basic Space Science, University of Nigeria, Nsukka, Nigeria*

[3] *Department of Mechanical Engineering, University of South Africa, Florida, South Africa*

[4] *Department of Engineering & Space Systems, National Space Research & Development Agency, Abuja, Nigeria*

Abstract: This chapter deals with the control devices used in automation such as Programmable Logic Devices (PLD), Programmable Logic Controller (PLC), Programmable Automation Controller (PAC), Personal Computer (PC), *etc.* The goal of the control devices in automation is to achieve an efficient, robust and reliable system control. Basically, system control devices include input devices (for raw data input), processing devices (for processing raw data into information), output devices (to disseminate the processed data and information), and storage devices (for the retention of processed data and information). The sensors feed the main controller with the input data acquired from the environment. Following the processing of the data, the decision is made by the main controller on the control action to take and this decision is communicated to the actuator for execution. The actuator in turn drives the final control device to implement the control action. The programming language is crucial in achieving optimum efficiency. While the PLC follows a scan-based program execution, PC software is usually event-driven. In terms of cost efficiency, indicators such as performance, expandability, and ruggedness are important considerations. The initial cost of a PC may be higher than that of a PLC as a PC is more suitable for processing of complex network loads. PLC may be initially inexpensive but as the demand for processing power increases, the PC-based system becomes more cost-effective. In terms of expandability, PLC usually offers support to standard industrial equipment but when an external control is needed, a PC is more suited. PLC does not require additional protection equipment compared to PC.

Keywords: PAC, PC, PLC, PLD.

* **Corresponding author Ilesanmi Afolabi Daniyan:** Department of Industrial Engineering, Tshwane University of Technology, Pretoria 0001, South Africa; Tel: +27 (064) 5298778; E-mail: afolabiilesanmi@yahoo.com

INTRODUCTION

Control devices are output devices that use input signals from sensors to change the state of a system or process. They can be in the form of mechanical, electronic or electro-mechanical devices and ranges from simple devices having a single loop such as a pump control to a sophisticated control such as the PLC having multiple inputs and outputs for full industrial automation. Unlike the single-loop controls, multi-loop controls can receive data from more than one sensor and provide control functions to more than one device.

PROGRAMMABLE LOGIC DEVICES (PLD)

Programmable Logic Devices (PLD) are integrated circuits that are programmed to perform different control functions according to the programs written in their memory. They usually employ low-level languages of commands and find applications in manufacturing automation. They consist of an array of the AND as well as the OR gates and the development software can be used to convert the basic codes into sets of instructions that a programmer can employ to implement the logic design.

The types of PLD types are classified as follows [1].

1. PROMs (Programmable Read Only Memory). This class offers high speed and low cost and it is suitable for relatively small designs.

2. PLAs (Programmable Logic Array): This class offers more flexible features which make it more suitable for complex designs.

3. PAL/GALs (Programmable Array Logic/Generic Array Logic): For this class, the random-logic gate networks are compactly laid out on an IC chip. This class offers good flexibility and speed and are less expensive than the PLAs.

The PLDs have several advantages and disadvantages depending on the type used.

For instance, the PLAs, like ROMs have the following advantages over PAL/GALs.

1. The design is not time-consuming compared to the PAL/GALs, characterized by time-consuming logic design of random-logic gate networks and an even more time-consuming layout.

2. The design is easy and flexible.

3. Quick and easy adoption of emerging technologies without the need for a change in the previous information.

3. Random-logic gate networks occupy smaller chip areas than PLAs or ROMs, although the logic design and the layout of random-logic gate networks are far more tedious and time- consuming.

4. Also, with large production volumes, random-logic gate networks are cheaper than PLAs or ROMs.

Compared to the ROMs, the PLAs have the following advantages. PLAs are smaller in size than the ROMs; this brings about a reduction in the board space requirement, and power requirement with improved compact circuitry. The disadvantage is that the merit of the small size of PLAs fizzles out as the number of terms in a disjunctive form increases. Thus, PLAs cannot store complex functions, most especially functions whose disjunctive forms consist of many product terms.

APPLICATIONS OF PLAS AND PAL/GALS

A micro-processor chip uses many PLAs because of the ease of design and flexibility. PLAs find applications in control logic and systems in which flexibility is a requirement. PLAs are used for code and micro-programs conversions, bus priority resolvers, decision tables and memory overlay.

PLA is also suitable for use when a new product is to be manufactured in small volume or as a test prototype due to the need for changes in the product at such stage. For a mature product that requires no further changes, the PLAs can be replaced by the random-logic gate networks for cost-effectiveness, volume production, and high speed.

A typical programmable machine has basic three components namely; a processor, memory as well as input and output devices.

Processor

This processes the information collected from the measurement system and takes logical decisions based on the information. The information is thereafter sent to the actuators or output devices.

Memory

The memory stores the input data collected from sensors and the programs to process the information in order to take necessary decisions or actions. A program is a set of instructions written for the processor to perform a specific task. A group of programs is called software.

Input/Output Devices

The input and output devices are used to communicate with the outside world or operator.

There are following three types of PLDs being employed in mechatronics systems; Programmable Logic Controllers, microprocessor, micro-controllers, and Proportional-Integral-Derivative (PID controller). The PLD has several advantages such as higher switching speed, compact circuitry, reduction in the power requirement, and reduction in the board space requirements.

The components of a PLD are shown in Fig. (**1**).

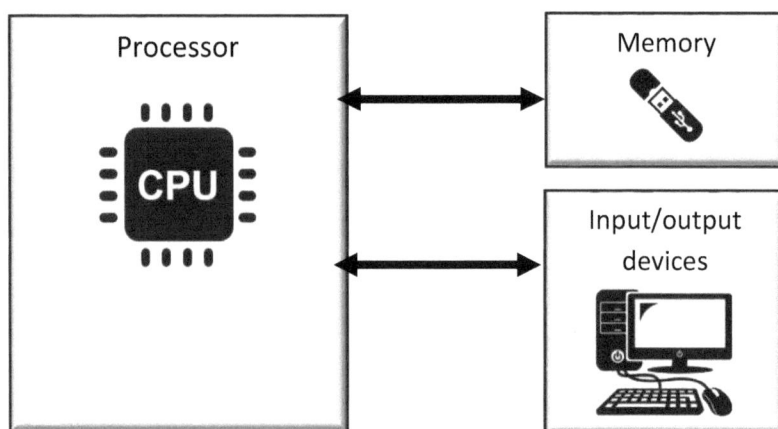

Fig. (1). Components of a PLD.

PROGRAMMABLE LOGIC CONTROLLER (PLC)

A programmable logic controller is a digital computer with a programmable memory usually employed for system's automation [2, 3]. It stores programs and implements functions such as logic, sequencing, timing, counting and arithmetic words in order to control machines and processes. PLCs are generally used for automating systems in an open loop where processes are to be performed in a sequential manner such as the automation of assembly lines in industries. Hence, it can automatically affect the control of various processes and conditions in an industrial system. It has a well-structured and programmed microprocessor as well as specially designed and implemented controllers which are unaffected by variations in the process conditions. When coding is programmed on a computer, it is transferred to the PLC where it is loaded and stored. The PLC has a non-volatile memory and effective storage capacity to retain programs. The PLC has input and output terminals. The input terminal interprets the logic states of the sensors and switches. In this case, the two logic states are HIGH or 1 as well as LOW or 0.

The PLCs are generally developed for systems with multiple inputs and multiple outputs (MIMO).

The following are some advantages of PLC;

1. They are cost-effective and easy to maintain.

2. High flexibility, easily programmable, and ability to use a similar system for other processes.

3. The programming interface is easier in comparison to other processers with adequate fault diagnosis.

4. They are highly reliable offering high resistance to impact and vibration.

5. They are also resistant to electrical and mechanical noise.

6. They are suitable for use at high temperatures.

7. They are suitable for operation control of electro-mechanical devices, especially in tough and hazardous industrial environments.

8. They are highly versatile as thousands of inputs and outputs can be controlled with a single PLC.

PLCs range from small devices having limited inputs and outputs (I/O), to large modular devices with a count of thousands of I/O which are networked to other PLCs and the SCADA system.

PLC finds application in process control, distributed control systems, networking, and in sequential control where all the operations are carried out in a sequence to automate a system such as the elevators, washing machine, traffic signals and automated vehicle assembly line.

MICROPROCESSOR

This is a silicon microchip device embedded with the Central Processing Unit of a computer system (CPU) for performing simple or complex arithmetic or logic operations. It is a multi-purpose digital integrated circuit that carries out necessary digital functions to process the information obtained from the measurement system. The microprocessor operates by reading or decoding a binary instruction from a storage device called the memory, and then processes the data according to the program in the binary form to produce an output. The microprocessor finds applications in reprogrammable systems such as the microcomputer and embedded systems such as the digital camera and photocopying machine. The advantage lies in its high speed of operation and its ability to quickly move data between various memory locations. For instance, a microprocessor with 5GHz can perform 5 billion tasks per second. The limitations, however, lie in the fact that it over-heats during operation and does not have internal peripherals like the ROM, RAM, and other I/O devices.

MICROCOMPUTER

A micro-computer uses a microprocessor as its central processing unit with memory, and a limited number of input and output devices such as keyboards, monitors, *etc.* which are integrated into a printed circuit board (PCB). It contains all functions of a computer and can take the decision that is used to control a system or actuate a course of action based on the input variables received. Standard microcomputers are laptops and desktops but generally, microcomputers may also include some mobile phones, workstations, and embedded systems. They also have memory in the form of read-only memory (ROM) and random access memory (RAM), input/output (I/O) ports, and a bus or system of interconnecting wires, all housed in a single unit referred to as the motherboard.

MICROCONTROLLER

It is a specific purpose control micro-processor based system that is dedicated and programmed for a specific control function in a system. A microcontroller may be programmed for the control of any process variable such as water level control, temperature, pressure, *etc.* It consists of a processor, memory, and programmable input and output peripherals and is basically designed for embedded and control applications.

The Arduino development boards are usually programmed with the use of an open-source integrated development environment (IDE) based on the processing programming language (usually written in C and C++). The IDE is users' friendly and includes a software library known as "Wiring" which simplifies common input and output operations. The Arduino programs in the IDE can be uploaded into the microcontroller through an existing boot loader. Although the higher abstraction influences the performance of the micro-controller chip negatively, the processor is still fast enough for most operations. The software functions are usually designed to measure the operation's or system's performance. Each performance measure is written in an algorithm and defined as a software function. The controller integrates the functionality of the other components through software algorithms to carry out the required functions of the system. Important selection criteria identified for a controller are that it should have sufficient memory for all the defined variables used in the software, a fast enough processor to carry out all the functions seamlessly, and it should be fitted with adequate general-purpose digital input and output pins to connect all the other required electrical components. Arduino technology has proven to be highly suitable as the main controller. The boards are cost-effective, readily available, easy to install, and supported by a large community on the Internet. Where the network connectivity requirements and onboard storage requirements are lacking, the Arduino can be combined with a networking shield. Some networking shields are combined with a micro SD storage module, which can serve as the storage function when the network is unavailable. The options for network connectivity can be Wi-Fi and Ethernet. Ethernet is more stable than Wi-Fi, and if it is close enough to connect to the network router, it would make a better choice. Connecting over Ethernet also provides the option to power the module over 'Power over Ethernet (PoE).

PROGRAMMABLE AUTOMATION CONTROLLER (PAC)

The PAC is a type of automation controller that incorporates higher-level instructions. PAC is often used in industrial control systems for the control of a wide range of systems and processes. The PAC can enable the provision of more sophisticated instructions to an automated system, thus, enabling much the same capabilities as PC-based controls in a single package like a PLC. The PAC boasts of a multi-domain functionality, simultaneous operation in multiple domains using a single platform (such as process control, motion control, sequential control, communication and data management), a single, multi-discipline development platform, flexible software tools which maximize process flow across processes, machines or systems, an open, modular architecture and compatibility with other networks. Hence, the PAC combines the functionality of a PLC with the processing capability of a PC.

CONCLUSION

It is noteworthy to mention that there is a difference between a controller and an actuator. A controller maintains the optimum performance of a system by introducing changes in its variable(s) so that the system can produce the desired output. On the other hand, an actuator is a device that implements changes that the controller introduces into the system. An actuator can be in the form of electric, hydraulic, pneumatic, mechanical, or electromechanical actuators depending on the system and its requirements. An actuator is part of the final control element that implements control actions based on the signal received from the controller. It is an energy source, which drives the final control element to act based on the different forms of energy applied.

The controller is an intermediate device, which issues commands in either the form of on/off, regulatory, stepwise, and in several forms to the actuator for implementation. The actuator in turn drives the final control device to implement the control action.

REFERENCES

[1] R.L. Shell, and E.L. Hall, *Handbook of Industrial Automation.* Marcel Dekker Inc.: New York, 2020.

[2] J.R. Hackworth, and F.D. Hackworth, *Programmable Logic Controllers: Programming Methods And Applications.* Pearson: New Jersey, 2004.

[3] M.P. Groover, *Automation, Production Systems, and Computer Integrated Manufacturing.* 3rd ed.. Prentice Hall, 2008.

CHAPTER 7

Industrial Automation Tools and Components

Ilesanmi Afolabi Daniyan[1,*], **Lanre Daniyan**[2], **Felix Ale**[3] and **Khumbulani Mpofu**[1]

[1] *Department of Industrial Engineering, Tshwane University of Technology, Pretoria 0001, South Africa*

[2] *Department of Instrumentation, Centre for Basic Space Science, University of Nigeria, Nsukka, Nigeria*

[3] *Department of Engineering & Space Systems, National Space Research & Development Agency, Abuja, Nigeria*

Abstract: Industrial automation tools are of a wide range of tools that are employed for industrial automation. These tools comprise various control systems that incorporate diverse sub-systems or devices to enhance industrial processes. Notable examples include Computer-aided design (CAD software) and Computer-aided manufacturing (CAM software), Artificial Neural Networks (ANN), Distributed Control Systems (DCS), Human-Machine Interface (MHI), Supervisory Control and Data Acquisition System (SCADA), instrumentation and robotics, *etc*. They are beneficiaries in the area of product development (improved design, analysis, and manufacture of products), quality control time and cost-effectiveness amongst others. This chapter emphasizes industrial automation tools and components. Furthermore, the application of industrial automation in robotics, packaging systems, computer numeric control systems, tool monitoring systems, advanced inspection systems as well as flexible manufacturing systems were discussed in this chapter. Industrial automation tools can significantly influence industrial processes with a reduction in manufacturing lead time, improvement in product quality, and effective process monitoring.

Keywords: ANN, Automation tools, CAD, CAM, DCS, HMI, SCADA.

INTRODUCTION

Industrial automation tools are tools that can be used to achieve process, system or industrial automation. They assist in creating mechanical or digital channels to solve manufacturing problems. These tools can also assist in achieving production efficiency as well as significant cost and time savings. This chapter discusses the types of industrial automation tools and their applications.

[*] **Corresponding author Ilesanmi Afolabi Daniyan:** Department of Industrial Engineering, Tshwane University of Technology, Pretoria 0001, South Africa; Tel: +27 (064) 5298778; E-mail: afolabiilesanmi@yahoo.com

Ilesanmi Afolabi Daniyan (Ed.)

TYPES OF INDUSTRIAL AUTOMATION TOOL

The following are the types of industrial automation tools;

1. The Artificial Neural Network (ANN)

2. The Distributed Control System (DCS)

3. The Human Machine Interface (HMI)

4. Supervisory Control and Data Acquisition (SCADA)

5. Programmable Logic Controller (PLC)

6. Instrumentation

7. Motion control

8. Robotics

The Artificial Neural Network (ANN)

An artificial neural network is a mathematical or computational model whose behaviour mimics those of biological neurons. The structure of the ANN is adaptive, as it can change based on the external or internal exchange of information within the network. Artificial neural networks are used to study, identify and correlate patterns in pools of data and to classify relationships (such as sequence recognition). Applications include e-mail spam filtering, system control (such as in a car), an automated irrigation system, pattern recognition in systems such as radars, pattern recognition in speech, movement, and text, and financial automated trading systems.

The Distributed Control System (DCS)

A distributed control system is the one that comprises separate controls throughout the system. The controls are not centrally located, but tend to be in the parts of the system that needs monitoring with each control connected to the others in a communication network. These kinds of systems are typically used in continuous manufacturing processes. For a given process, the controllers can be specified and manipulated to enhance or monitor machine performance. The distributed control system finds application in oil refining and central station power generation as well as in robotic operations. For instance, the traffic lights are usually controlled by a distributed control system.

Human-Machine Interfaces (HMI)

Human-machine interfaces (HMI) or computer human interfaces (CHI) are usually used to communicate with PLCs and other computers [1]. A human machine interface system depends on human interaction with the system in order to function with the user who will provide the input while the output that coincides with the user's intent will be produced by the system. For example, the automated teller machine (ATM) is designed so users can easily manipulate the system thereby providing the appropriate results. The buttons on the machine provide the user an avenue to trigger a chain of commands within the internal system.

Supervisory Control and Data Acquisition System (SCADA)

A supervisory control and data acquisition system (SCADA) is a larger, industrial control network often comprising smaller sub-systems, including human machine interface systems that are connected to remote terminal units. The system translate sensor signals into a comprehensible data and the SCADA system can effectively control an entire manufacturing site, by linking up the different manufacturing plants in the site [2]. The term SCADA is normally associated with the development of control systems that cover a large geographic area. A SCADA is used for centralized monitoring and control of field sites over long-distance communication networks with the field devices employed for local operations such as the collection of data from sensor, opening and closing of valves and breakers, as well as the monitoring of the entire environment. SCADA systems are similar to distributed control systems, and the key difference lies in their function. SCADA systems do not control each process in real time, rather they coordinate processes.

Instrumentation

Instrumentation refers to the collection of instruments used for indicating, measuring, and recording physical quantities [1]. It also refers to the ability to monitor or measure the level of a product's performance, to diagnose errors in order to rectify them where necessary. The operation of measuring instruments is used in the design and configuration of automated systems in electrical, hydraulic and pneumatic domains as well as in the control of quantities being measured. Instrumentation finds application in the measurement of physical values of process parameters such as viscosity, flow, temperature, density, liquid level, pressure, voltage, inductance, resistance, capacitance, vibration, weight, current, frequency, *etc*.

Robotics

Robotics deals with the design, construction and operation of robots in conjunction with computer systems for control, information processing and feedback. Robots on the other hand are reprogrammable multifunctional machines that perform repetitive tasks that are dangerous, stressful, boring and intensive for human *via* sets of computer programs or instructions [3]. Robotic technology finds application in industrial and manufacturing processes *e.g.* welding, painting, machine tending, machining, assembly work, material handling, *etc*.

COMPONENTS OF AN INDUSTRIAL AUTOMATION SYSTEM

An industrial automation system requires the following components [1, 2].

1. Production system or plant to be used for production,

2. Sensors and actuators,

3. Communication system (for communication with the production system and with the automating computer systems),

4. The computer system comprising of the hard ware (computer hardware and process periphery), and the software (application/user and system's software).

The software is a set of all programs necessary for the execution of an automation task. The software program is divided into two types namely; the application and the system's software. Application software inputs the measured values and calculates the control variables while the system's software is the main driver of the operating system. The application software is necessary for the acquisition of process variables as well as process control and monitoring. On the other hand, the system's software is a run-time compilation program for organising the application program, data traffic, and enhancing effective communication between the operator and the computer. It is also necessary to manage peripheral devices.

APPLICATIONS OF INDUSTRIAL AUTOMATION

Industrial automation finds application in the following areas amongst others:

1. Robotics

2. Packaging systems

3. Computer Numerical Control Machines

4. Tool monitoring system

5. Advanced system inspection

6. Flexible manufacturing system

Robotics

Robots are programmable machines which receive signals from the system and environment to carry out programmed activities autonomously or semi autonomously. They take decisions and interact with other interfaces as well as the central control system *via* the sensors and actuators. Robots combine the techniques of numerical control and remote control to replace human workers with numerically controlled mechanical actuators [3].

Robots are classified into two categories; Artificially Intelligent robots and Non-Artificially intelligent robots.

Artificially Intelligent Robots

These are robots which are controlled by artificial intelligence programs. They require artificial intelligence algorithms involving learning, perception, problem-solving, language-understanding and logical reasoning to perform complex tasks. Most times the artificial intelligence involves some level of machine learning, where an algorithm is "trained" to respond to a particular input in a certain way using known inputs and outputs. For operational flexibility, robots are supplied with sophisticated techniques of feedback, vision and tactile sensors, reasoning capabilities, and adaptive control. The artificially intelligent robots are incorporated with devices such as the machine's vision comprising an array of sensors and cameras which increase their learning, and perception. Thus, improving their capacity to interact the environment and make decision. They can also communicate effectively with humans and carry out computational analysis in real time. They may also possess the playback and feedback facilities. However, the artificially intelligent robots usually require a high level of computing and computer control, as well as advanced programming language in order to input the decision-making logic and other artificial intelligence programs into their memory. For example artificial intelligence can be used to train a robot's vision to detect the kind of objects it's interfacing with.

Non-artificially Intelligent Robots

They simply carry out a defined sequence of instructions without the use of artificial intelligent programs. Many robots are not artificially intelligent. Most industrial robots are programmed to carry out a repetitive series of movements. Repetitive movements do not require artificial intelligence. An example of a non-artificial intelligent robot is the collaborative robot that is used to perform the autonomous work of picking up an object and placing it elsewhere. The work is repetitive without any need for artificial intelligence.

There are generally two types of sensors used in robotics. They are sensors for internal purposes and for external purposes. Internal sensors are used to monitor and control the various joints of the robot thereby forming a feedback control loop with the robot controller. Examples of internal sensors include potentiometers, optical encoders, tachometers of various types deployed to control the speed of the robot arm.

On the other hand, external sensors are external to the robot itself, and are used to control the operations of the robot. External sensors are simple devices that determine the correct positioning of a part or whether a part is ready for loading or unloading. An example of this is limit switches.

The areas of robotic applications in manufacturing include the following;

1. Material/Part processing; Robots are employed for processing some work piece parts in some production, assembly or manufacturing processes such as welding, forging, riveting, drilling, grinding, finishing operations, *etc.*

2. Material/Part handling; These involve activities such as sorting, part separation, part picking and placement as well as loading and unloading of machine parts *e.g.* in heat treatment operations, die casting and press work.

3. Product development and inspection; robots find important application in subassembly and assembly of products during product development such as the assembly of automobile parts and other joining processes such as bolting, riveting, welding, bonding, *etc.*

Robots are often employed in any of the following work situations;

i. Hazardous environment: Robots can be employed for operation in an unsafe, unhealthy or hazardous environment for humans or for operations in some areas that are inaccessible to human personnel.

ii. Specialized work cycle: When the same series or sequence of operation is required, robots can perform work with repeatability with the high degree of precision.

iii. Difficult work handling: Where some part are difficult or intensive to handle for humans during assembly operations, robots can handle such parts effectively.

iv. Multi-shift operations: A robot can be used in place of many operators in many work shifts with the high degree of efficiency.

v. Long Production runs: Robots find important use in long production runs such as in continuous operations since it is not prone to fatigue due to long term use.

Packaging System

Automatic packing of products is not only efficient but also ideal for consumable foods as chances of contamination *via* human handling are greatly reduced.

Computer Numeric Control Systems

These are machines operated and controlled by computer programs. There are limitations in the rate of production using conventional machine tools such as the lathe, milling, drilling machine *etc.* because time is wasted during tool, work piece and machining conditions set up as well as during job loading and unloading. The CNC system automatically sets the machining parameters using some codes of instructions consisting of numbers, letters of the alphabets, and symbols which the machine control unit (MCU) understands. These instructions otherwise known as programs are converted into electrical pulses which are followed by the machine's motors to perform the required operations on a work piece. The specific distances, positions, functions or motions of the tool and work piece are represented by the coded set of instructions denoted by numbers, letters of the alphabets, and symbols. CNC automatically guides the axial movements of machine tools with the help of computer control. Other auxiliary operations such as coolant on-off, tool change, and door open-close are automated with the help of micro-controllers.

Tool Monitoring System

It is important to monitor the cutting tool for wear and breakage during cutting operations in order to prevent time wastage, tool and work piece damage and fatalities. Complete automation of a machining process predicts when there is a tool wear during the course of machining operation. These systems are often employed for the prediction of tool breakage or wear and give alarms to the system operator to prevent catastrophic failure or damage of the machine or cutting tool and the work piece. Conventional tool monitoring systems rely on temperature of cutting, machining time, number of usage, torque, power, electrical resistances *etc.* to determine tool wear.

Advanced Inspection System

The test for the product integrity, quality, uniformity and conformability of a product is essential before the product is released into the market to avoid negative feedback that can ruin the integrity of the organization. In order to save time and improve the process of product inspection, product inspection process is automated. These technologies encompass the various sensory and data acquisition systems as well as machine vision systems and metrology instruments such as the co-ordinate measuring machine (CMM), digital calipers and screw gauges. Nowadays, quality control activities are being carried out right from the initial design stage of product development to ensure strict adherence to production plans and prompt correction in cases where deviations exist. Various physical and experimental-based simulation and modelling software are used to predict the real performance of the product or the system before development.

Flexible Manufacturing System

A flexible manufacturing system (FMS) for manufacturing is a system which integrates the numerically-controlled machine tools with automated material handling to ensure the flow of component parts, measuring equipment and a central computer control for product development with less human intervention. The system's flexibility permits automatic change over along different process routes and sequences of operations, allowing automatic machining of different products in small batches in the same system. Centralized FMS have often proved too complex, and they are subdivided into smaller flexible manufacturing cells (FMC) that include several CNC machines, robots, and transfer devices controlled by a single computer, the "cell controller.

CONCLUSION

This chapter discussed industrial automation tools and components. The different types of industrial automation tools such as Artificial Neural Network (ANN), Distributed Control Systems (DCS), Human-Machine Interface (MHI), Supervisory Control and Data Acquisition System (SCADA), instrumentation and robotics have been highlighted. Furthermore, the application of industrial automation in robotics, packaging systems, computer numeric control systems, tool monitoring systems, advanced inspection systems as well as flexible manufacturing systems is discussed.

REFERENCES

[1]   R.L. Shell, and E.L. Hall, *Handbook of Industrial Automation*. Marcel Dekker Inc.: New York, 2020.

[2]   "Industrial Automation. The IDC Engineers and Bookboon.com", Available at: http://155.0.32.9:8080/jspui/bitstream/123456789/486/1/Industrial%20Automation.pdf Accessed 30th May, 2022.

[3]   "JM608: Industrial Automation. Port Dickson Politeknik", Available at: https://nikarifblog.files.
         wordpress.com/2017/12/jm608-industriial-automation-textbook.pdf Accessed 30th May, 2022.

CHAPTER 8

Practical Examples of System Automation and Control

Ilesanmi Afolabi Daniyan[1,*], **Lanre Daniyan**[2], **Adefemi Adeodu**[3] and **Felix Ale**[4]

[1] *Department of Industrial Engineering, Tshwane University of Technology, Pretoria 0001, South Africa*

[2] *Department of Instrumentation, Centre for Basic Space Science, University of Nigeria, Nsukka, Nigeria*

[3] *Department of Mechanical Engineering, University of South Africa, Florida, South Africa*

[4] *Department of Engineering & Space Systems, National Space Research & Development Agency, Abuja, Nigeria*

Abstract: This chapter provides a practical demonstration of how a system's automation can be achieved. Some specific examples presented include the automation of irrigation systems, waste segregator, gasifiers, biodiesel plants, biogas plants, lawn mowers, assembly line automation as well as the automation and control of the suspension system of a railcar. The details of the design and components required for the automation of these systems are highlighted. The chapter presents practical guided approaches by which system automation can be achieved depending on the end-users requirements. The practical examples highlight the integration of sensors (for measuring conditions/parameters), controllers (for processing inputs and decision-making) as well as actuators (for effecting changes) with minimal or no human interference. System automation is connected to the engineering field called mechatronics which is an interdisciplinary engineering branch comprising a combination of mechanical, computer, electrical and electronic systems. The automation of systems will enhance profitability, improved production rate, product quality, and safety.

Keywords: Actuator, Controller, Sensor, System automation.

INTRODUCTION

The preceding chapters have dealt with the theoretical concepts of automation and control. This chapter discusses the practical ways by which processes or systems can be automated and controlled. It is noteworthy to mention that there is no single solution or approach for the automation and control of all systems as the

---

* **Corresponding author Ilesanmi Afolabi Daniyan:** Department of Industrial Engineering, Tshwane University of Technology, Pretoria 0001, South Africa; Tel: +27 (064) 5298778; E-mail: afolabiilesanmi@yahoo.com

automation and control of production systems can be tailored towards the production requirements and the bottom-line goal of profitability. This chapter discusses various systems as well as the materials required for their automation and control including feasible methods by which this can be achieved.

IRRIGATION SYSTEM

The purpose of automating the irrigation system is to reduce human intervention in supplying water for farmland irrigation and also to enhance effective water management. The capability of an automated irrigation system is the ability to take intelligent decisions with respect to the need for irrigation based on operational and soil conditions [1, 2].

AUTOMATION OF THE IRRIGATION SYSTEM

The automation of a standalone irrigation system will require the following; an Arduino microcontroller (main controller), an oscillator, a DC water pump, a PV module, LCD, a real-time clock module, a soil moisture sensor, a sim card, switches, relay, an led, capacitors and resistors [1, 2]. The Arduino Uno microcontroller provides a modular way of interfacing the real world with computer systems to get some activities accomplished on a chip in connection with the sensors. The Arduino whose technical specification is shown in Table **1** will be sufficient for the automation of a mini-standalone irrigation system.

Table 1. Technical specifications of the Arduino Uno.

S/N	Item	Value	Remarks
1	Micro-controller	8-bit Atmel ATmega328p	Arduino
2	Operational voltage	5V	The range of input is 6-12 V
3	Digital GPIO	14	6 capable of PWM
4	Analog IO	6	10-bit
5	Program memory	Flash 32kb, EEPROM 1kb	SRAM 2kb
6	Clock speed	16MHz	-
7	USB	Type B socket	-
8	Programmer	In-system firmware	USB-based
9	Serial communications	SPI, I2C	Software UART
10	Other	RTC, watchdog, interrupts	-

The Arduino is programmable with the use of Arduino IDE coded in a programming language.

System Architecture

The soil moisture sensor will check for the environmental conditions with respect to the pre-set time and feeds the microcontroller. The microcontroller is programmed such that the actuator is actuated for irrigation once the soil moisture level falls below 25%. The microcontroller also stops the actuator once the soil moisture rises up to 50%. Anytime any of these two conditions is met the microcontroller activates the relay to switch on the irrigation pump for irrigation purposes. On the microcontroller, the signal is read, and convertedinto a digital signal with the threshold humidity preset on it. Anytime these two conditions are active (time and humidity value falling below the threshold), a relay is activated by the microcontroller to switch the irrigation pump ON. A Short Message Service (SMS) alert is then sent to the authorized user on the state of the pump and the active mode of the control. The status of the pump changes from "PUMP OFF" to "PUMP ON" once the relay activates the pump. In case of an emergency need for water that is not dependent on the soil moisture, the system is incorporated with a manual mini switch which can be used to activate the pump.

Fig. (**1**) presents the architecture of the developed irrigation system.

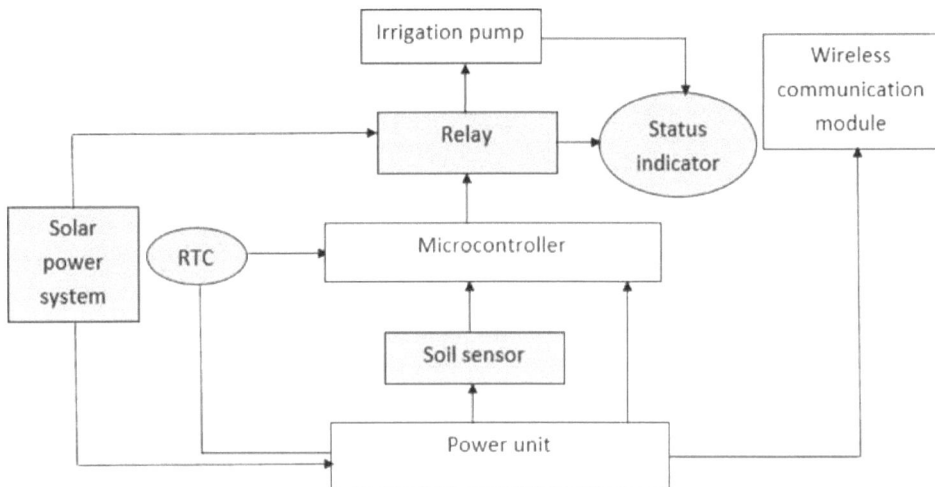

Fig. (1). Architecture of the developed automated irrigation system [1].

A timer (DS1307) is integrated to precisely record and track the time and date of the irrigation operation.

Fig. (**2**) presents the Proteus diagram showing the integration of the components of the irrigation system.

The irrigation system can work both onsite and offsite. It can work remotely *via* a Short Message System (SMS) sent *via* a mobile device to an authorized user. The data gathered by the sensor on the soil conditions can also be stored on the cloud storage or repository and can be shared in real time with the aid of Internet of Things (IoT) devices. This makes it easy for data analytics to analyze the data acquired for the prediction on the soil condition as well as the time and date suitable for the irrigation purpose. An authorized user can get a periodic update on the soil conditions *via* SMS notifications. This enables such a user to plan irrigation schedules and respond swiftly to irrigation demands.

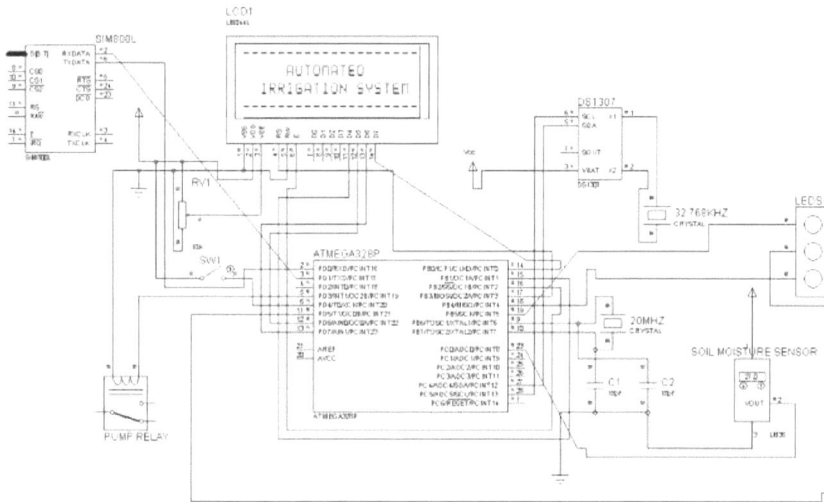

Fig. (2). Proteus diagram showing the integration of the components of the irrigation system [2].

Waste Segregator

The aim of the automated waste segregator is to enhance the automatic detection and sorting of waste with minimal or without human interference. The capacitive proximity sensor as well as the ultrasonic sensor are required for easy detection and perception. Other components required include a central microcontroller for implementing control actions, a vero and circuit boards, as well as a motor driver for motion actuation. The major components required in the electrical and electronic phase are the capacitive proximity sensor for the detection of the type of material to be sorted (either plastics or metals) based on the differences in the magnitude of their dielectric constant. The ultrasonic sensor and a DC geared motor with high speed and torque features can be coupled to the lids of the segregator for automatic lid opening and closing. The microcontroller controls the whole process of system's segregation in real-time. The microcontroller serves a contact point where all other systems' peripheral devices are linked together. On the circuit and vero boards, the electrical and other system components are connected using connecting wires. The motor driver can be used for regulating the speed of the electric DC motors. Other accessories may include: battery, pushbuttons, resistors and capacitors, voltage regulator, oscillator, as well as integrated circuit chip sockets.

The programming was done using a suitable programming language usually the (C programming language). The implementation of the system comprises three major phases namely: motor control, measurement using the sensor and object's distance measurement with the aid of an ultrasonic sensor. A DC motor is employed to drive the wastes towards the capacitive proximity sensor and voltages will be induced as a result of change in the electric field. The various lids of containers for different waste categories depend on the voltage readings obtained from the capacitive proximity and ultrasonic sensors. In a situation whereby the ultrasonic sensor does not give a distance reading *vis-à-vis* the pre-set value on the micro controller, the motor will remain in the OFF state. On the other hand, when a certain distance is detected, the microcontroller reads the voltage indicated by the capacitive or proximity sensor. Once a voltage is read by the microcontroller, the microcontroller will open the corresponding lid of the container for either the plastic, metals or other waste lids and the waste sensed will be deposited.

Gasifier

A gasifier is a system that can convert solid fuels such as wood chips into safe gaseous fuels [3, 4]. The components required for the automation of a small scale downward gasifier include the main controller (Arduino micro controller), which has the minimum and maximum threshold values of the temperature, pressure and flow rate was pre-set on it [3, 4]. A thermocouple is used for sensing the temperature, a pressure transducer for pressure measurement and calibrated flow rate sensors for measuring the flow rate respectively. The output of the process parameter measured will be amplified using an amplifier and then converted from analogue to digital using the Analogue to Digital Converter on the micro controller. The micro controller adjusts the process conditions in real time based on the comparators' feedback between the measured parameters and the threshold.

The programming of the Arduino board can be done using a programming language and the codes communicated with the Android mobile for monitoring, sending queries and receiving feedbacks.

Biodiesel Plant

A biodiesel plant is a plant for the conversion of renewable biological resources such as vegetable oil or animal fat into a methyl or ethyl ester *via* the process of transesterification [5, 6].

The automation of a small to medium-scale biodiesel plant will involve the use of a thermocouple/temperature probe, connected to the heating element for temperature measurement, a pressure transducer, for pressure measurement, a flow sensor for flow rate measurement, *etc.* depending on the critical process parameters which can significantly affect the production process that needs to be effectively monitored [7, 8]. Measured signals from the sensors are passed on to the amplifier connected to the output for voltage amplification. The amplified voltage will serve as the output of the amplifier and then be passed through a microcontroller where the analogue signal is converted into a digital signal using Analogue to Digital Converter (ADC) on the micro controller. On the controller, the threshold process parameters (temperature, pressure, flow rate) are pre-set to the preferred range for all the compartments. The voltage measured is compared with the threshold value pre-set on the microcontroller. If the value of the process parameter measured exceeds the threshold, the microcontroller will simply turn OFF the contactor or relay, and the component such as the heater, or pumps, *etc.* will be automatically deactivated. The Arduino microcontroller serves as the main controller while the contactor or relay, which activates or deactivates components such as the heater, pump, *etc.* is the actuator. The micro controller can be linked to an internal timer that is connected to a buzzer, to send an alarm through a configured switch (transistor circuit). The progress of the measurement as well as the monitoring and control actions are displayed on the Liquid Crystal Display (LCD).

The Pressure Transducer

This measures the pressure of the system with a sensing material and converts measured values into voltage. It consists of an elastic material that deforms under the application of pressure and an electrical element, which detects the deformation and transmits it as changes in voltage. The amplifier is employed to amplify the differential output voltage between the output pin 4 and 1 of the pressure sensor. The output of the amplifier is coupled to the analog pin (A0) of the Arduino microcontroller and thereafter, it is converted into digital input voltage, which is displayed on the LCD.

The pressure transducers can be mounted on the system and interfaced together. The positive and negative terminals of the pressure transducer are connected to the power source while the transducer output (yellow wire) is connected to its respective plant input terminal.

Temperature Sensor

The temperature sensor is a device capable of measuring temperature *via* an electric signal. It also has the capability to monitor the temperature of the liquid raw materials. The probe is usually made of stainless steel material for good corrosion resistance.

The Timing System

The production processing activities usually involve the presetting of the time required for a certain operation. The pre-programmed time for these activities is set and stored on the micro controller. Hence, the microcontroller routinely checks for this time with respect to other process conditions. A timing system also known as the real-time clock can be incorporated to give accurate track of date and time.

Float Switch

This is a type of sensor used for the detection of the liquid level in a tank. The switching device with a rated power, has a resistance output signal in ohms and a range of operating temperatures. The output of the float switch is usually in the analog form and the value is converted on the microcontroller into digital with the aid of the analog-to-digital converter (ADC). The hardware components include: the Arduino, sensor, connecting wires and pin configurations.

Flow Meter

The flow meter is a device employed for the measurement of the volumetric flow rate of the fluid *via* the pipe orifice. The flow meter which is of the positive displacement type usually has a liquid temperature in the range. It also has the capability to maintain steady accuracy when the viscosities change.

Buzzer

The buzzer is an electromechanical audio signaling device, which produces an alarm when a crucial control action is needed

LM6009 and LM 2596 Modules

Both the LM6009 and LM 2596 modules are used for voltage regulation. The LM 6009 module is a step-up, non-isolated voltage converter. It has a switching current of 4 V with an adaptable input voltage in the range of 4.2-32 V as well as an output voltage range of 5-52 V at high operational efficiency. On the other hand, the LM 2596 is a step-down converter with a constant and adaptable input voltage within the range of 1.25-35 V DC while the output voltage is in the range of 4-40 V DC. It has an output current of 2 A and a frequency of 150 kHz.

The Wireless Communication and SMS Module

The wireless communication module comprises two wireless modules, a USB serial adapter, an active sim and SIM 800L, GSM/GPRS module for sending and receiving messages.

Leakage Detection in the Pipeline

This system incorporates an AP-40 sensor with 31/2-digit, 2-colour, 7-segment LED with a character height of 11 mm, a display cycle of 5 times/cycle and a rated pressure of -15 to 110% of F.S for leakage detection. The AP-40 is a sensor, which is specifically designed to detect the physical presence of oil and pressure leakages in flow lines. As shown in the flowchart (Fig. **3**), the microcontroller upon reception of a signal from AP-40 of product leakage, instructs the buzzer to sound an alarm based on the connection of the component on the printed circuit board assembly (PCBA).

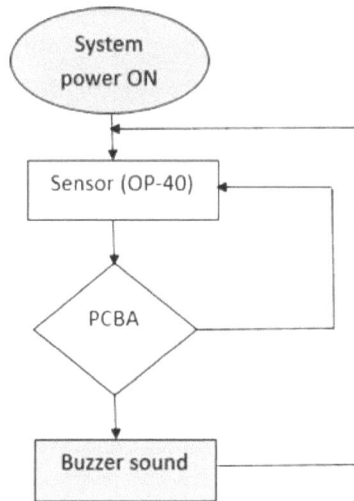

Fig. (3). Flow chart for leakage detection in pipes.

Biogas Plant

A biogas plant is a plant for the production of a combustible mix of gases produced by the natural fermentation of wet biomass in an anaerobic process [8,

9]. The components for the automation include: Arduino Uno (ATmega328), a pressure transducer, a temperature sensor and a pH meter kit [8 - 11]. The pressure transducer measures the pressure of gas while the temperature sensor is for temperature detection. The pressure transducer may have a working pressure range of 0-1.2 MPa. The normal working temperature range is 0-85°C and the response time is approximately 2 ms.

The pH meter kit is a kit that measures the pH of a substance. It is specially designed for the Arduino and has an accuracy of ± 0.1 pH (at 25°C). The kit has a range of 0 – 14pH. The kit consists of a pH sensor probe, a BNC connector and a pH 2.0 interface.

Lawn Mower

A lawn mower is a machine that employs a revolving blade to cut a lawn to an even height [12]. The design of the automatic lawn mowing machine encompasses its obstacle-avoiding ability and performance of the mowing operation with the least human interference. The major sub-systems include a power supply unit, an IR led / receiver sensor pair, an ultrasonic sensor, an Arduino microcontroller and the geared DC motors. The microcontroller powers the sensors which are used in obstacle avoidance and path planning of the geared DC motors. The IR LED/ receiver unit is made up of an IR LED and an IR receiver placed in parallel to each other. While the IR LED emits light in the forward direction, the IR receiver would receive the reflected IR light. If there is an obstacle ahead of the lawn mower, the ultrasonic sensor will provide the path planning which in this case is a helix path from the end-end across the lawn until it covers the entire lawn. This is achieved by the ultrasonic sensor which detects the pavement of the boundaries of the lawn by producing a sound that is reflected back and received by its piezo speaker. This information is transmitted together with the range of obstacles to the microcontroller. The motor takes its power supply from the batteries connected in series to form a power source. The microcontroller sends a series of pulses from its PWM pins which determine how fast the motor would rotate. All sub-systems are linked together. The automation of this robot requires the integration of the micro controller, ultrasonic sensors, and motor driver.

At the front, the ultrasonic sensors detect any obstacle ahead and send a signal to the microcontroller for accurate control and steering of the motors. The microcontroller and the motor driver provide the needed control of the motors. The micro IDE software can be used to compile the programs with the code downloaded to the microcontroller. Programs can be written for each sensor separately to interface them with the microcontroller. Once the shaft encoder, the proximity sensor, the compass and the motor controller are interfaced with the microcontroller, a GPS coordinating map can be used to identify the border of the field while the system's vision can be used to locate the boundary lines and provide feedback necessary for line following [12].

Alert System for Human Access Control

The materials necessary for the development of a small control alert system for human access are shown in Table **3**.

Table 3. Materials for control alert development.

S/N	Materials	Description	Quantity
1	Micro-controller	8-bit Atmel ATmega328p	1
2	Transformer	220- v- 12 v	1
3	Relay	1 k	1
4	Full bridge rectifier	-	3
5	Key pad	-	1
6	Vero-board	-	3
7	Bread board	-	3
8	Finger print scanner	-	1
9	L.C.D	-	4
10	Transistor	L7812	1
11	Sliding door	-	1

Arduino Uno Microcontroller

The Arduino microcontroller (ATmega328p with 14 digital input/output pins) comprises the mixed signal type integrating analog components to perform non-digital functions. 6 of the pins can be used as PWM output while 6 are used as analog inputs. Other accessories may include a 16 MHZ quartz crystal, a USB connection, a power jack, as well as an ICSP header and a reset button. It contains everything needed to support the microcontroller. It is simply connected to a computer with a USB cable or powered with an AC-to-DC adapter or battery to get started.

A simple way to power the Arduino board is *via* a USB connection or with an external power supply. The power source is selected automatically. It can also be powered externally either with an AC-to –DC adapter, or with the use of a battery. It should be ensured that the board operates within the safe operating range of voltage according to specifications. When the voltage supply exceeds the specification, the regulator may over heat and damage the board. The power pins are as follows;

1. VIN; the input voltage to the Arduino board when it uses an external power source (as opposed to the voltage from the USB connection or other regulated power source).

2. GND; Ground pins.

3. IOREF; This pin on the Arduino board provides the voltage reference with which the microcontroller operates. The description of the Arduino Uno is presented in Table **4.**

Table 4. Technical specifications of the Arduino Uno.

S/N	Item	Value	Remarks
1	Micro-controller	8-bit Atmel ATmega328p	1 mm sheet metal
2	Operational voltage	5 V	Input range: 7-12 V
3	Digital GPIO	14	6 capable of PWM
4	Analog IO	6	10-bit
5	Program memory	Flash 32 kB, EEPROM 1kb	SRAM 2 kB
6	Clock speed	16MHz	-
7	USB	Type B socket	-
8	Programmer	In-system firmware	USB-based
9	Serial communications	SPI, I2C	Software UART
10	Other	RTC, watchdog, interrupts	-

The Alert System

Fig. (**4**) presents the integration of the various components employed for the development of the remote controlled system.

The alert system comprises three major systems namely: the power source, transmission system, and the reception system. The reception system is further divided into three main systems namely the alert, actuation and display system.

The transmission and reception systems are powered by the power source. A direct current (DC) power source is employed for powering the transmission system while an alternating current (AC) power source is used for powering the reception, alert and actuation systems.

Fig. (**4**): Generalized block diagram for remote controlled alert system for automobile access.

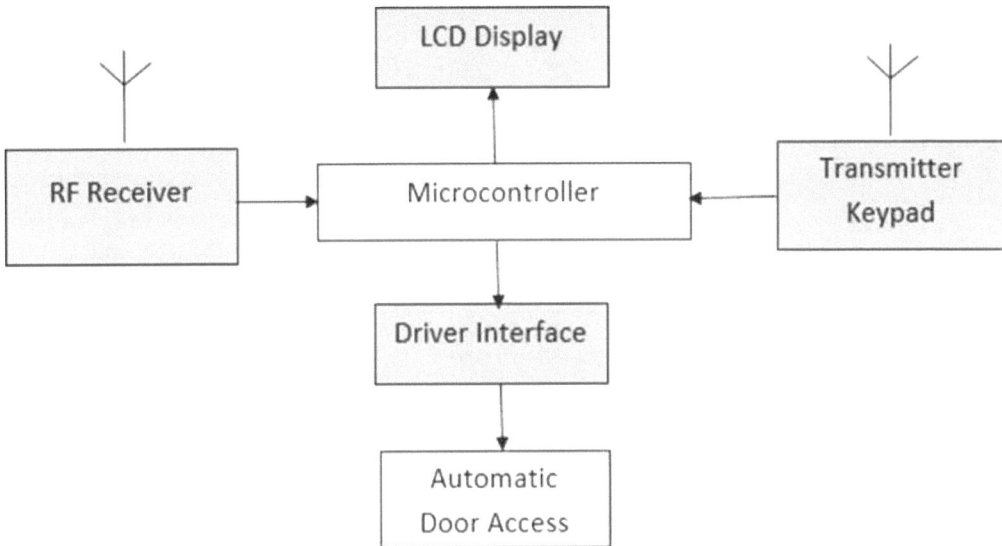

Fig. (4). Flow diagram for the access control.

Reception System

The reception converts the transmitted signals into an audio signal and kinetic energy. The components of the reception system: Antenna, RLP 315 RF receiver, 220V AC, ATMEGA8 microcontroller, LCD, transformer, rectifier, 7805V voltage regulator, 1K resistor, buzzer, 470µF capacitor, 220µF capacitor (filter), 100Ω resistor, and HT12D RF decoder.

The Alert System

The alert system converts the output from the decoder into an audio signal *via* the control function of the micro controller on the transistor to which the audio device is connected. The buzzer is employed for giving the audio signal while a bipolar transistor acts as a switch for the buzzer. The transistor is connected to a designated port which generates commands in programmed pulses with the aid of a 1K resistor. Thus, the timed pulses are converted into timed audio signals.

The Actuation System

The function of the actuation system is to convert the source from the designated port in the microcontroller into a sliding motion. The torque servo and the mechanism of the sliding door enable the conversion of the electronic signal into physical motion. The servo motor is powered by the low current source of the PORT D from the ATMEGA8 microcontroller which causes its rotor to rotate at 0.04A, at an angle of 180°, on a dead zone of 2 microseconds and a torque of 9.4kg/cm (Jerry, 2001). This motion is converted into linear motion by the sliding door mechanism which causes the door to slide 180° forward, as an "open" action or 180° backward, as a "close" action for vehicle access into the premises. The sliding door mechanism consists of two rectangular cuts of wood of length 300 mm, groove depth of 4mm, groove width of 2mm and 8mm thickness; a model gate of the length of 200 mm and a width of 400 mm, with a rectangular slut of 180mm and width of 10.5mm, to which a slut pin of length 6mm, and a diameter of 10mm is attached, with the pin cap of thickness 10 mm and diameter 15 mm, which is also connected to the circumference of a circular plate of diameter 100 mm, thickness of 10 mm, having a hole of 2 mm diameter at its centre, and a groove of 4mm diameter and a depth of 6mm also at its centre.

The tower pro MG996R metal gear digital torque servo rotates 180° according to the ATMEGA8 program after the security pin is typed on the remote control, causing the circular plate to rotate the slut pin; and subsequently, a push motion is initiated to the sliding door to which it is connected thus leading to circular to linear motion conversion. Later on, the tower pro MG996R metal gear digital torque servo rotates 180° in reverse motion according to the ATMEGA8 program, causing a "pull" action that closes the model gate.

The Proteus simulation testing is shown in Fig. (**5**).

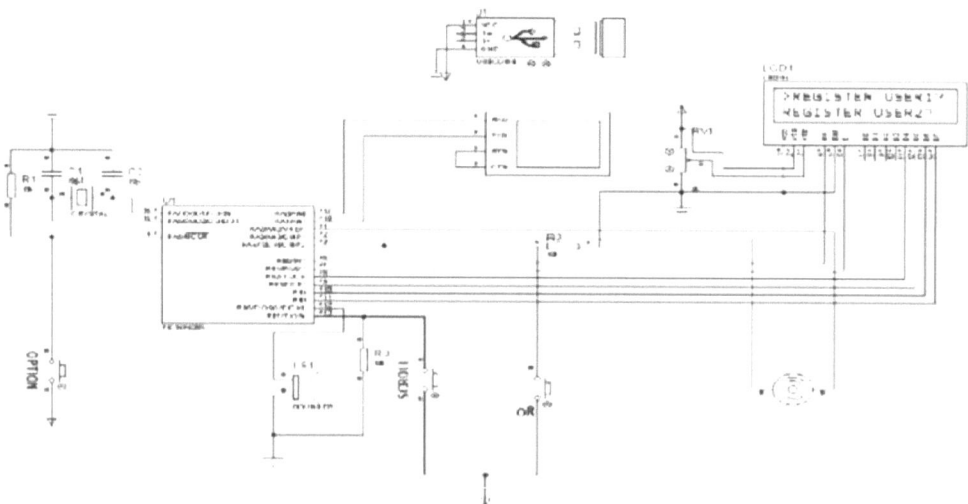

Fig. (5). Proteus diagram for human access into the premise.

AUTOMATION OF ASSEMBLY LINE

The aim of an automated assembly line is to minimise assembly errors, automate monotonous or repetitive tasks, and provide high productivity and efficiency of operation. This work proposes an intelligent system for assembly operations that can be applied in the automotive or rail sector. The intelligent system consists of a CCD camera, an array of smart sensors, a part component box, an LED light source, a control system, a conveyor as well as a robot for material handling and assembly. The component parts to be assembled are classified and stored according to sizes into standard and non-standard parts. Fig. (**6**) shows the Proteus diagram of the intelligent system for the assembly operation. It comprises two precision stepper motors (Robots A and B), one for pick and place operations and the other for assembly operations. Below each of the stepper motors is their drivers (L296H –Bridge). At the centre is the main controller (ATMEG) with the written code of the sequence of the assembly operations. D1 and D2 are diode LEDs to indicate the status of the component movements. D1 turns ON when robot A is placing a component on the conveyor. This implies that robot A is active while placing and turns OFF once it has placed the object. D2, on the other hand, turns ON when robot B is in operation (when the component has arrived for assembly operation). A trans-receiver sensor is placed in between robots A and B to transmit a code to alert robot B to wait to receive a component once D1 turns ON and vice versa. This is a coordinating robot that directs root A to pick and place components on the conveyor and alert B to be ready to receive it.

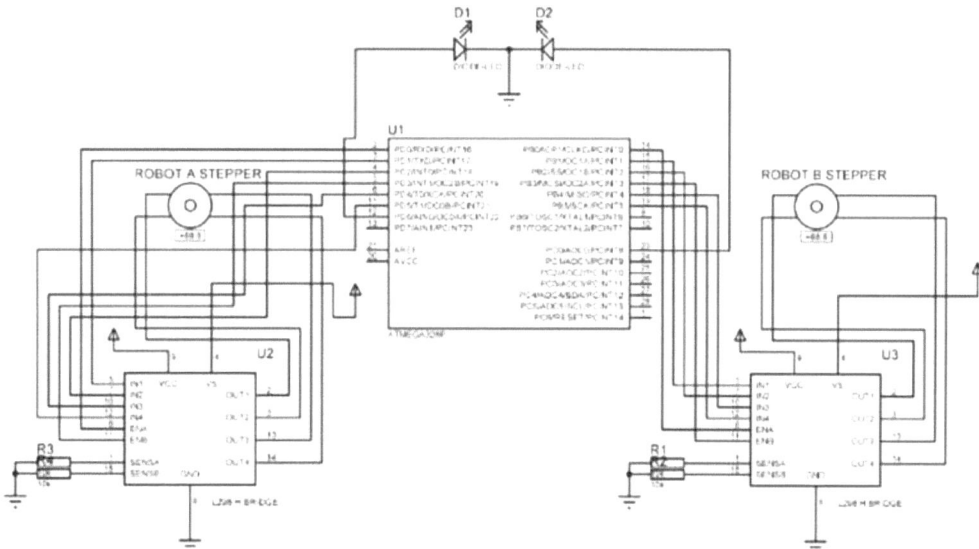

Fig. (6). The Proteus diagram of the robot circuitry.

AUTOMATIC CONTROL OF THE SUSPENSION OF A RAILCAR

The automatic control of the railcar suspension system is geared towards achieving significant ride comfort during the movement of the railcar. The movement of the railcar is characterized by vibration due to irregular rail profiles as well as load disturbances as a result of the load carried by the railcar. These vibrations and disturbances can offset the balance of the railcar system thereby causing ride discomfort. Thus the primary and secondary suspension systems of the railcar can be controlled such that they will respond and adjust in real-time to cancel out the effect of the disturbances. The suspension system of a railcar comprises the railcar bogie, wheelset, primary and secondary suspension systems, side frame, bolster, wheelset, axle, bearing adapter and roller bearings. The primary suspension system connects the rail track to the railcar bogie while the secondary suspension system connects the railcar bogie to the railcar body. The suspension system can be passively or actively controlled depending on the level of automation and the system's performance requirements. The passive suspension system makes use of the spring and damper for storing energy and dissipating vibration energy while the active suspension system employs the sensors and actuators for adjusting the acceleration, speed and displacement of the suspension system during movement. The control of the suspension systems can be achieved with the use of classic (for instance the PID control), advanced or adaptive controls (for instance, the fuzzy PID) depending on the level of control or automation desired.

Fig. (7) shows the free-body diagram for the model of the suspension system of a railcar. The model comprises the sprung mass denoted as ms and the unspring mass denoted as mu. The suspension system consists of a passive spring Fks and a damper Fbs in parallel with an active control force F.

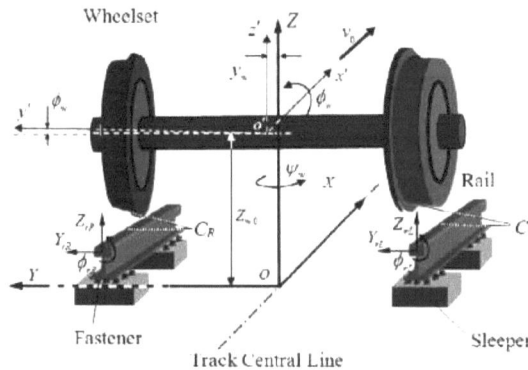

Fig. (7). The free-body diagram for the rail car model [13].

From Newton's second law of motion and the free body diagram shown in Fig. (**7**), the dynamic equation was obtained thus;

Let,

$$F_{m_s} = m_s \ddot{x}_1 \tag{1}$$

$$F_{m_u} = m_u \ddot{x}_2 \tag{2}$$

$$F_{k_s} = k_s^l (x_2 - x_1) + k_s^{nl} (x_2 - x_1)^3 \tag{3}$$

$$F_{k_t} = k_t (x_2 - w) \tag{4}$$

$$F_{b_s} = b_s^l (\dot{x}_2 - \dot{x}_1) \tag{5}$$

Let $b_t = 0$ then, $F_{bt} = 0$

$$+\uparrow \sum F = m\ddot{x} \tag{6}$$

Considering m_s

This implies that,

$$F_{m_s} = F_{K_S} - F + F_{b_s} \tag{7}$$

$$m_s \ddot{x}_1 = k_s^l (x_2 - x_1) + k_s^{nl} (x_2 - x_1)^3 - F + b_s^l (\dot{x}_2 - \dot{x}_1) \tag{8}$$

$$\ddot{x}_1 = \frac{1}{m_s} [k_s^l (x_2 - x_1) + k_s^{nl} (x_2 - x_1)^3 - F + b_s^l (\dot{x}_2 - \dot{x}_1)] \tag{9}$$

Considering m_u

$$F_{m_u} = -F_{k_s} + F - F_{b_s} + F_{k_t} \tag{10}$$

$$m_u \ddot{x}_2 = -k_s^l (x_2 - x_1) - k_s^{nl} (x_2 - x_1)^3 + F - b_s^l (\dot{x}_2 - \dot{x}_1) + k_t (x_2 - w) \tag{11}$$

$$\ddot{x}_2 = \frac{1}{m_u} [-k_s^l (x_2 - x_1) - k_s^{nl} (x_2 - x_1)^3 + F - b_s^l (\dot{x}_2 - \dot{x}_1) + k_t (x_2 - w)] \tag{12}$$

For the state space representation,

Let:

$$\dot{x}_1 = x_3 \tag{13}$$

$$\dot{x}_2 = x_4 \tag{14}$$

$$\dot{x}_3 = \ddot{x}_1 = \frac{1}{m_s}[k_s^l(x_2 - x_1) + k_s^{nl}(x_2 - x_1)^3 - F + b_s^l(\dot{x}_2 - \dot{x}_1)] \tag{15}$$

$$\dot{x}_4 = \ddot{x}_2 = \frac{1}{m_u}[-k_s^l(x_2 - x_1) - k_s^{nl}(x_2 - x_1)^3 + F - b_s^l(\dot{x}_2 - \dot{x}_1) + k_t(x_2 - w) \tag{16}$$

Using the state space representation and neglecting the non-linear components in order to obtain the linear time-invariant state space representation, Eq. (17) holds thus,

$$\dot{x}(t) = fx(t) + gu(t) \tag{17}$$

Eqs. (18) and (19) respectively express state, input, and output vector.

$$x = [x_1, x_2, x_3, x_4 \ldots]^T \tag{18}$$

The system's input u(t) is the input into the linear dynamic system having output y(t) with respect to the current state x(t) Eq. (19).

$$y(t) = Cx(t) + Du(t) \tag{19}$$

The model of the railcar comprises of the following Cartesian coordinates: vertical (a_1, a_2, a_3), longitudinal (l_1, l_2, l_3), lateral directions (m_1, m_2, m_3), bogie and track (roll, pitch and yaw motions) including their angles respectively: ϕ_x, ϕ_y, ϕ_{x1}, ϕ_{y1}, ϕ_{x2}, ϕ_{y2}.

The vertical displacement in vector form is expressed as Eq. (20).

$$a_j = \{a_1, a_2, a_3, l_1, l_2, l_3, m_1, m_2, m_3, \varphi_x, \varphi_y, \varphi_{x1}, \varphi_{y1}, \varphi_{x2}, \varphi_{y2}\}^T \tag{20}$$

The suspension system is designed such that it will be able to isolate disturbances by being soft against rail disturbances and hard on load disturbances. A transfer function $\left(\frac{x}{y}\right)$ which is the ratio of the steady state vibration response (x) to the frequency of steady state motion of the rail disturbances (y) is introduced.

Thus, Eqs. (21) and (22) hold as follows:

$$H_{x/y}(\omega) = \frac{x}{y}e^{-i\varnothing} \tag{21}$$

Where;

$$\left|\frac{x}{y}\right| = \frac{[1+[2\tau\frac{\omega}{\omega_n}]^2]^{\frac{1}{2}}}{[(1-\frac{\omega^2}{\omega^2})^2+[2\tau\frac{\omega}{\omega_n}]^2]^{\frac{1}{2}}} \tag{22}$$

The phase angle is expresses by Eq. (23).

$$\varnothing = tan^{-1}\frac{[2\tau\frac{\omega}{\omega_n}]^2}{1-\frac{\omega}{\omega_n}+[2\tau\frac{\omega}{\omega_n}]^2} \tag{23}$$

The response $x(t)$ is written as Eq. (24).

$$x(t) = |y||H_{x/y}(\omega)|\cos(\omega t - \varnothing) \tag{24}$$

To isolate rail vibrations from the rail car body, the ratio of the operating frequency of the rail disturbance (ω) to the natural frequency of the rail car system (ω_n) must be greater than $\sqrt{2}$ as expressed in Eqs. (25) and (26).

$$\frac{\omega}{\omega_n} > \sqrt{2} = 1.1414 \tag{25}$$

$$\text{Hence } \left|\frac{x}{y}\right| < 1 \tag{26}$$

Where; y_w is the lateral displacement (m), ψ_m is the yawing angle (deg.), ϕ_w is the rolling angle (deg.), $Y_{rL,R}$ is the lateral displacement along the horizontal axis (m), $Z_{rL,R}$ is the vertical displacement (m), $\phi_{rL,R}$ is the torsional angle (deg.), X, Y, *and* Z are the system axes with coordinates x^l, y^l and z^l.

m_s is the sprung mass (kg); m_u is the unspring mass (kg), k_s^l is the linear component of the spring constant (N/m), k_s^{nl} is the nonlinear component of the spring constant (N/m), k_t is the wheel spring constant (N/m), b_s^l is the linear component of the damping coefficient (Ns/m), b_s^{nl} is the nonlinear component of the damping coefficient, (Ns/m), $y = x_2 - x_1$ is the suspension deflection (m), x_1 is the body deflection (m), x_2 is the wheel deflection (m), w is the road disturbance (m), u is the system input, $F = Ax_p$ is the actuator force (N), F_{ks} is the spring force acting on the body (N), F_{bs} is the damping force acting on the body (N) and F_{kt} is the spring force acting on the wheel (N).

The following model parameters presented in Table **5** were used for simulation.

Table 5. Model parameters for the control of the railcar suspension system [14, 15].

S/N	Parameter	Notation	Value	Unit
1.	The average mass of the rail car	Mr	50,500	Kg
2.	The average mass of bogie	Mb	2,410	kg
3.	Mass of primary suspension system	Mp	30,000	kg
4.	Mass of secondary suspension system	Ms	30,000	kg
5.	Moments of inertia	Ii	56900	kgm²
6.	Rail car roll inertia	Ir	68,200	kgm²
7.	Rail car pitch inertia	Ip	71,000	kgm²
8.	Average mass of first wheelset and axle	m1	1300	kg
9.	Average mass of second wheel	m2	1,300	kg
10.	Distance between the centre of gravity and the front position of the rail car	di	6	m
11.	Distance between the centre of gravity and the middle position of the rail car	d2	6	m
12.	Distance between the centre of gravity and the rear position of the rail car	d3	6	m
13.	Spring constant of the primary suspension system	k1	2.4×10⁶	N/m
14.	Spring constant of the secondary suspension system	k2	5.6×10⁵	N/m
15..	Spring constant of the wheel	k3	4.0×10⁵	N/m
16..	Damping constant of the primary suspension system	b1	1.2×10³	Ns/m
17.	Damping constant of the secondary suspension system	b2	2.95×10⁴	Ns/m
18.	Damping constant of the wheel	b3	5.0×10⁴	Ns/m

Hence, the output of the system written in matrix form is expressed by Eq. (27).

$$Z(t) = Ax(t) + By(t) + Cz(t) \qquad (27)$$

Considering the control input U(S), the transfer function for the primary suspension system is expressed as Eqs. (28) and (29).

$$G_1(s) = \frac{X_1(s) - X_2(s)}{U(s)} \tag{28}$$

$$G_1(s) = \frac{(M_r + M_p)*s^2 + b_2*s + k_2}{((M_r*s^2 + b_1*s + k_1)*(M_p*s^2 + (b_1 + b_2)*s(k_1 + k_2)) - (b_1*s + k_1)*(b_1*s + k_1))} \tag{29}$$

In addition, considering the rail disturbance input W(S), the transfer function is expressed as Eqs. (30) and (31).

$$G_2(s) = \frac{X_1(s) - X_2(s)}{W(s)} \tag{30}$$

$$G_2(s) = \frac{(-M_r + b_2*s^3 - M_r + k_2*s^2)}{((M_r*s^2 + b_1*s + k_1)*(M_p*s^2 + (b_1 + b_2)*s(k_1 + k_2)) - (b_1*s + k_1)*(b_1*s + k_1))} \tag{31}$$

On the other hand, the transfer function for the secondary suspension system is expressed as Eqs. (32) and (33).

$$G_1(s) = \frac{(M_r + M_s)*s^2 + b_3*s + k_3}{((M_r*s^2 + b_2*s + k_2)*(M_s*s^2 + (b_2 + b_3)*s(k_2 + k_3)) - (b_2*s + k_2)*(b_2*s + k_2))} \tag{32}$$

$$G_2(s) = \frac{(-M_r + b_3*s^3 - M_r + k_3*s^2)}{((M_r*s^2 + b_2*s + k_2)*(M_r*s^2 + (b_2 + b_3)*s(k_2 + k_3)) - (b_2*s + k_2)*(b_2*s + k_2))} \tag{33}$$

Fig. (**8**) shows the architecture of the adaptive control (Fuzzy-PID) involving the use of Fuzzy logic to control the PID. The adaptive control is geared towards the control of the suspension system of the railcar.

The following are some of the steps necessary to control the suspension system of the railcar.

1. Formulation of the mathematical model for the railcar and its suspension system including the degree of freedom and the possible types of motion during movement.

2. Generating of equation of motions.

3. Modelling of the railcar and its suspension systems as well as the rail track.

4. Simulation of the modelled railcar and its suspension systems as well as the rail track. This can be carried out in the MATLAB-Simulink.

5. Controller design and the introduction of disturbance rejection control.

6. Adjustment of the controls until the desired performance is achieved.

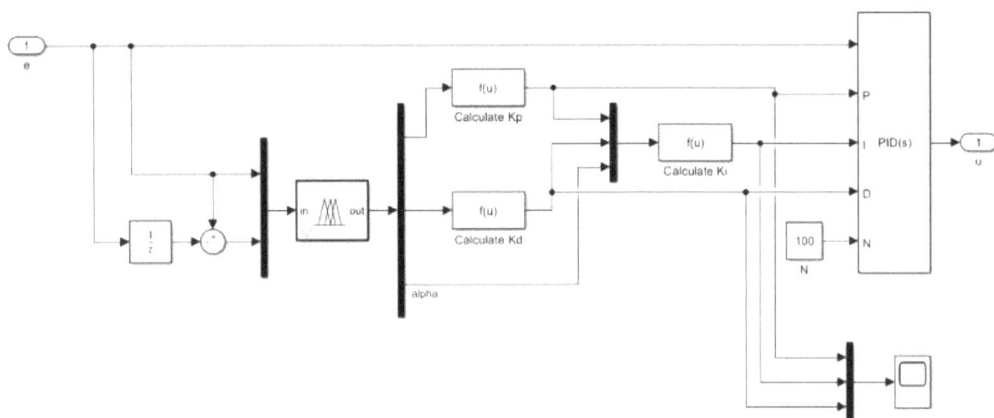

Fig. (8). Architecture of the Fuzzy-PID for the control of the railcar suspension system.

Fig. (**9**) presents the Bode diagram for the control of the railcar suspension system. The figure shows how the railcar system exhibits different types of damping: underdamping, overdamping, critically damped and underdamping. At the underdamped state, the system oscillates through the equilibrium position and causes only a small change in the behavior of an oscillatory system. Thus, the frequency of oscillation decreases by only a small amount while the amplitude decays gradually with time. As shown in the Figure, the damping coefficient at the underdamped state is less than 1. For the over damped system, the system moves more slowly toward equilibrium compared to the system that is critically damped. The damping coefficient for the underdamped system is greater than 1. For the critical damping, the system returns to the equilibrium position quickly without an overshoot and the damping coefficient for the critically damped system equals 1. The underdamped system will oscillate through the equilibrium position for a long time without energy loss. At the undamped state, the damping coefficient equals zero.

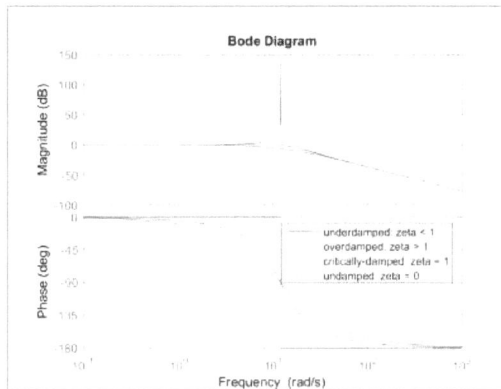

Fig. (9). The Bode diagram for the control of the railcar suspension system.

Existing works have reported that the use of classic, advance and adaptive controls can minimise the rise and settling time and eliminate steady state errors. The active controls can also minimise the amplitude of oscillations and overshoot in real time thus providing stability and improvement of the rail holding ability of the suspension system [14 - 16].

CONCLUSION

This chapter highlights some practical examples of system's automation; the design concept, materials for development and performance evaluation. Some specific examples presented include: automation of irrigation system, waste segregator, gasifier, biodiesel plant, biogas plant, lawn mower, assembly line automation as well as the automation and control of the suspension system of a railcar. The details of the design and components required for the automation of these systems are highlighted.

REFERENCES

[1]   I.A. Daniyan, O.L. Daniyan, K. Mpofu, and R. Boitumelu, and B. I. Ramatsetse Development and performance evaluation of automated irrigation system. Proceedings of the 2019 IEEE SAUPEC/RobMech/PRASA Conference Bloemfontein, South Africa, January 28-30, 29. Added to IEEE Xplore, 978-1-7281-0369-3/19. pp. 12-16, 2019.
[http://dx.doi.org/10.1109/RoboMech.2019.8704785]

[2]   L. Daniyan, E. Nwachukwu, I. Daniyan, O. Bonaventure, I. A. Daniyan, and B. Okere, "Development and optimization of an automated irrigation system", *Journal of Automation, Mobile Robotics and Intelligent Systems,* vol. 13, no. 1, pp. 37-45, 2019.
[http://dx.doi.org/10.14313/JAMRIS_1-2019/5]

[3]   I.A. Daniyan, K. Mpofu, A.O. Adeodu, and O. Momoh, "Development of a 12 kW-capacity gasifier for domestic and experimental applications", *Proceedings of the 2020 IEEE Southern African Universities Power Engineering Conference/Robotics and Mechatronics/Pattern Recognition Association of South Africa.* Added to IEEE Xplore, pp. 68-73, 2020.

[http://dx.doi.org/10.1109/SAUPEC/RobMech/PRASA48453.2020.9040963]

[4]   I. Daniyan, F. Ale, I.D. Uchegbu, K. Bello, and M. Osazele, "Process optimization and performance evaluation of a downdraft gasifier for energy generation from wood biomass", *Acta Polytech.,* vol. 61, no. 5, pp. 601-616, 2021.
[http://dx.doi.org/10.14311/AP.2021.61.0601]

[5]   I. Daniyan, E. Bello, T. Ogedengbe, and K. Mpofu, "Use of central composite design and artificial neural network for predicting the yield of biodiesel", *Procedia CIRP,* vol. 89, pp. 59-67, 2020.
[http://dx.doi.org/10.1016/j.procir.2020.05.119]

[6]   I.A. Daniyan, E.I. Bello, T.I. Ogedengbe, and P.B. Mogaji, "Gas chromatography and fourier transform infrared analysis of biodiesel from used and unused palm olein oil", *International Journal of Engineering Research in Africa,* vol. 42, pp. 47-64, 2019.
[http://dx.doi.org/10.4028/www.scientific.net/JERA.42.47]

[7]   I.A. Daniyan, K. Mpofu, O.L. Daniyan, and A.O. Adeodu, "Development and Automation of a Multi-feedstock Plant for Biodiesel Production", *Proceedings of the 2021 IEEE International Conference on Electrical, Computer and Energy Technologies (ICECET),* pp. 1-9, 2021.
[http://dx.doi.org/10.1109/ICECET52533.2021.9698675]

[8]   I.A. Daniyan, O.L. Daniyan, A.O. Adeodu, I.D. Uchegbu, and K. Mpofu, "Performance evaluation of smart multi feedstock biodiesel plant", *Procedia Manuf.,* vol. 35, pp. 1117-1122, 2019.
[http://dx.doi.org/10.1016/j.promfg.2019.06.065]

[9]   I.A. Daniyan, O.L. Daniyan, O.H. Abiona, and K. Mpofu, "Development and optimization of a smart system for the production of biogas using poultry and pig dung", *Procedia Manuf.,* vol. 35, pp. 1190-1195, 2019.
[http://dx.doi.org/10.1016/j.promfg.2019.06.076]

[10]   I.A. Daniyan, A.K. Ahwin, A.A. Aderoba, and O.L. Daniyan, "Development of a smart digester for the production of biogas", *Petrol. Coal,* vol. 60, no. 5, pp. 804-821, 2018.

[11]   I.A. Daniyan, A.K. Ahwin, A. K., A.A. Aderoba, O.L. Daniyan, and K. Mpofu, "Development and optimization of a smart digester for the production of biogas", *Proceedings of the International Conference on Industrial Engineering and Operations Management Pretoria* Johannesburg, South Africa, pp. 1456-1459, 2018.
[http://dx.doi.org/10.1016/j.promfg.2019.06.076]

[12]   I. Daniyan, V. Balogun, A. Adeodu, B. Oladapo, J.K. Peter, and K. Mpofu, "A. O., B. I. Oladapo, J. K. Peter and K. Mpofu. Development and performance evaluation of a robot for lawn mowing", *Procedia Manuf.,* vol. 49, pp. 42-48, 2020.
[http://dx.doi.org/10.1016/j.promfg.2020.06.009]

[13]   L. Ling, X-B. Xiao, J-Y. Xiong, L. Zhou, Z-F. Wen, and X-S. Jin, "A three-dimensional model for coupling dynamics analysis of high speed train-track system", *Journal of Applied Physics and Engineering,* vol. 21, p. 1, 2014.

[14]   I.A. Daniyan, and K. Mpofu, "K. and D. F. Osadare. Design and simulation of a controller for an active suspension system of a rail car", *Cogent Eng.,* vol. 5, no. 1545409, pp. 1-15, 2018.

[15]   I. A. Daniyan, K. Mpofu, O. L. Daniyan, A. O. Adeodu, and A. O. Dynamic Modelling, *Cogent Eng.,* vol. 6, no. 1602927, pp. 1-20, 2019.

[16]   I.A. Daniyan, K. Mpofu, A.O. Adeodu, and O.L. Daniyan, "Model Design, Simulation and Control of Rail car Suspension System", *Published Proceedings of the 2019 IEEE 10th International Conference on Mechanical and Intelligent Manufacturing Technologies (ICMIMT 2019)* Cape Town, South Africa IEEE, pp. 37-42, 2019.
[http://dx.doi.org/10.1109/ICMIMT.2019.8712066]

CHAPTER 9

Water Distribution Management in Real Time: Using a Cloud-Based Approach

Kazeem Aderemi Bello[1,*], Ilesanmi Afolabi Daniyan[2], Osato Alexendra Ighodaro[3], Adefemi Adeodu[3] and **Wasiu Adeyemi Oke[4]**

[1] *Department of Mechanical Engineering, Federal University, Oye-Ekiti, Nigeria*

[2] *Department of Industrial Engineering, Tshwane University of Technology, Pretoria0001, South Africa*

[3] *Department of Mechanical and Mechatronics, Afe Babalola University, Ado Ekiti, Nigeria*

[4] *Department of Mechanical Engineering, University of South Africa, Florida South Africa*

Abstract: Lack of access to potable water has become an issue of concern in our society. In order to satisfy the increasing water demands of the galloping population in Nigerian communities, it is essential to use smart technology to manage water resources. The purpose of this research work is to ensure that inaccessibility to water as a result of pump failure is detected in real-time through smart technology. To solve this daunting challenge, an Arduino microcontroller and Liquid Crystal Display (LCD) were used to switch on and switch off the submersible pump at a predetermined Water Level (WL) in the tank and also to determine the pump availability hours. The WL in the tank was monitored using an Arduino microcontroller, sensors, relays, and LCD. A Global System for Mobile telecommunication (GSM) module was also used to create an interactive medium between the user/maintenance team and the system to monitor the submersible pump reliability based on engineering theory and concept. The system was tested by introducing varying volumes of water in a constructed water distribution system prototype in the laboratory. The microcontroller was efficient in controlling the system; the pump was able to switch on and switch off when the WL in the tank was 50% and 100%, respectively. As an autonomous system, it was capable of taking decisions automatically without human interference. The system was able to send feedback *via* SMS to alert the user/maintenance team to check the pump whenever it failed to pump water at WL$\leq$ 50%. This innovative design system will help to monitor and manage water distribution properly and it should be considered for use in schools, hospitals, residences, offices, *etc*. to ensure the availability of water always, save energy consumption and help in combating covid-19.

Keywords: GSM module and Arduino microcontroller, Real-time, Submersible pump, Water distribution.

* **Corresponding authors Kazeem Aderemi Bello:** Department of Mechanical Engineering, Federal University, Oye-Ekiti, Nigeria. Tel: +234 (080) 36386760; E-mail: itanooluwaponmile@yahoo.com

Ilesanmi Afolabi Daniyan (Ed.)
All rights reserved-© 2023 Bentham Science Publishers

INTRODUCTION

Lack of access to potable water in Nigerian communities has been a great challenge, although there is no global water scarcity, accessibility has been a major challenge [1]. The purpose of this research work is to ensure that inaccessibility to water as a result of pump failure is detected in real-time through the use of smart technology. The sustainability of available water resources in many parts of the world is now a dominant issue [2]. This problem is about poor water allocation, inefficient use, and lack of adequate and integrated water management [3]. The common method of level control in water storage tanks for most Nigerian communities is simply to start the pump when the water level in the storage tank is low and allow it to run until a higher water level is reached or the water overflows from the water storage tank. This method is not adequate for proper control of the system. Liquid level control systems are widely used for monitoring liquid levels. Usually, this kind of system provides continuous level indication [4]. Proper water distribution and management are required to ensure water sustainability. Presently, floaters are being used to regulate the pumping of water. This does not provide an efficient method of monitoring the water in the tank. The need to constantly wash hands as a way of combating COVID-19 has further necessitated easy accessibility to water in many homes, offices, colleges, industries, *etc.* in recent times. The challenge of lack of access to water may not be due to a lack of resources to make water available but most times it happens due to faults in the system that go undetected for a long time. In residential areas, people switch ON the water pumps and set off to work or fall asleep, forgetting to switch OFF the water pump when the tank is full. This results in wastage of water due to overflow in overhead tanks, flooding of premises, and wastage of energy. Without proper water monitoring and management, water cannot be available and supplied properly.

Clean water is not accessible to a significant percentage of the world's population [5]. It is estimated that about 2.2 billion people worldwide lack access to potable water and about 297,000 children less than five years of age die every year from unsafe drinking water [6]. Water-related diseases are the single largest cause of human sickness and death in the world and mostly affect the poor population [7]. The UN estimates that each person requires at least 50 liters to 100 liters of clean, safe water a day, for drinking, cooking, washing, bathing, flushing toilets, planting, sanitation, *etc* [8]. Although clean water is scarce all over the world, Africa is the world's poorest and most underdeveloped continent because of a myriad of reasons that embrace a lack of access to potable water [9]. Water scarcity trends as one of the foremost underestimated problems globally, it is one of the more threatening challenges in Africa [10]. Water scarcity is due to physical shortage, or scarcity in access which itself is a result of failure to ensure regular supply or because of lack of infrastructure to make clean water accessible. The Food and Agriculture Organization [11] estimates that by 2030, up to 250 million individuals in Africa would be living in areas of high-water crisis. Due to a lack of sanitation and purification strategies, many people in Ghana suffer from water-related diseases [12]. About 80% of wealthy Nigerians have access to minimum basic water supply, compared to only 48% of poor Nigerians [13]. Nigeria's three tiers of government share the responsibility for managing water resources and for providing water, but over the years, there has been gross inefficiency and confusion in the area of water distribution and management. The available rural wells are fitted either with submersible pumps or with hand-operated pumps for water supply. These sources yield very little or no water throughout the dry season and are at risk of frequent breakdowns, resulting in water shortages and even crises. Usually, after only a few years, several of those wells become faulty due to lack of funds for operation or poor maintenance, sometimes, general maintenance is nonexistent [14]. A more flexible and responsible water management system is required. In a research work by Olalekan [15], the researcher observed that the water crisis in Nigeria is not new; it has been embedded in the Nigerian States since the colonial days.

In advanced countries, there is a new technology for "atmospheric water generation" that provides high-quality water by extracting water from the air by cooling the air into condensed water vapour [16]. A submersible pump driven by an electrical motor raises the water to the surface. Most times, a deep well may actually penetrate a confined aquifer; in this case, hydrostatic pressure will raise the water to the surface naturally. Drilling deep prevents microorganisms like bacteria and pollution from entering the well. This also reduces issues of wells drying up because of seasonal fluctuations (the distinction between wet and dry seasons) in the water table. The process of pumping water from underground is very expensive hence restricting the full development and use of all groundwater resources [17].

Surface water and groundwater are both important sources of community water supply needs. Groundwater is a common source for homes and small towns [17]. These systems consist of pipes, pumps, valves, storage tanks, reservoirs, meters, fittings, public hydrants, chlorination equipment and other hydraulic accessories. According to Faizah [18], the design of a distribution system depends on the determination of storage, the location and size of the feeder, the location and size of distribution pipes, valves, and hydrants, and the determination of the pressure required in the system. In designing a water distribution system, it is necessary to survey the area leading to the source of supply. Often a trained engineer carries out the required survey, to determine the route that the feeder pipe will follow and the location of distributing reservoirs and main pipes. This survey is followed by a topographic survey to determine the locations and elevations of all low and high points within the area. The choice of the type of distribution system depends on three factors: topography, location, and size of the distribution area, elevation and site layout. There are three types of water distribution networks based on the method of distribution: gravity system, direct pumped system and combination system [18]. According to the National Research Council [19], a good distribution system is expected to convey potable water from the source to consumers with the same level of purity. The system should be able to deliver the water to the consumer at the required pressure and quantity. The use of adequate pipe quantity and good quality minimizes water leakage during distribution. The distribution system is required to be economical and easy to operate and maintain. Also, it should be safe from future pollution.

The recently developed technology for monitoring and management features is the Internet of Things (IoT) technology. The IoT is growing rapidly in the developed world. Several authors have defined the term "Internet of Things" differently. For instance, Pallavi and Smruti [20] defined the Internet of Things as "a paradigm in which objects equipped with sensors, actuators, and processors communicate with each other to serve a meaningful purpose". IoT is not a single technology but a combination of various technologies working together. The application of IoT in different aspects of our daily activities has helped in reducing human intervention and preserving energy and resources like water. In the water supply and distribution system, researchers are developing smart solutions using IoT technology to manage water resources efficiently [21]. Several researchers and developers have made modifications and given proposals on how to improve water availability and accessibility with the use of technology such as the Internet of Things. In a recent research by Odiagbe *et al.* [22], the authors suggested the use of Internet of Things technology to monitor water quantity and pipeline leakage in domestic water distribution networks to ensure the supply of potable drinking water. They developed a system capable of detecting water leaks. The system can control and restrict the flow of water during leak detection in the main water distribution system. The authors made use of a solenoid valve to regulate the flow of water; a microcontroller activates the valve whenever there is leak detection. A flow sensor was used to measure the flow rate of water through the pipeline. In a paper by Nunes *et al.* [23], they designed an internet of things-based platform for real-time management of energy consumption in water resource recovery for wastewater. The authors discussed the design and implementation of adequate meters for measuring different electrical parameters (including energy consumption). For the pilot demonstration, the authors made use of Wi-Fi to send information to the communication server and later used LoRa (long-range) communication technology. Other researchers have introduced IoT-based technology in the agricultural sector in order to boost productivity. Daniyan *et al.* [24] and Daniyan *et al.* [25] reported on the development and automation of an automated irrigation system. The control and automation of the standalone irrigation system were done using an AVR microcontroller, programmed to activate an intelligent and independent farm irrigation operation *via* a water pump attached to the system. Morteza *et al.* [26] developed a system referred to as a "multi-intelligent control system (MICS)" of a water pump and a pump station for the agricultural sector application. The system consisted of three control systems including the electro-pump controller, water level in the reservoir and alarm control system. The whole system consisted of five smart layouts, including smart communication, security, monitoring, alert and smart water level, which were all operated by the Internet of Things technology thus making the results accessible from anywhere. The authors found the MICS to be reliable and convenient in agricultural and industrial sectors as well as for domestic applications. They also reported an increase in the efficiency and productivity of the water management system by 60%. In a research by Kaminski *et al.* [27], a Smart Water Management Platform (SWMP) design was introduced for agricultural applications using IoT-based technology for precision irrigation. They applied a hands-on approach based on four pilots and the outcome showed high performance for the SWMP pilots. However, good design configuration of some components to give maximum scalability by applying reduced compu-tational resources will be necessary. In a paper by Cristina *et al.* [28], the authors proposed an efficient distributed monitoring and control approach for a water utility to reduce the current water loss. The proposed approach was to help utility operators improve water management systems by exploiting new technologies like the Internet of Things. The authors proposed an IoT-based architecture for water utility monitoring and control. The architecture was divided into the following levels: sensor level, communication, management and application, terminal and user level. According to Cristina *et al.* [28], the internet of things could be one of the most efficient approaches for developing more utility-proper systems to manage water resources. Maruthi *et al.* [29] developed an IoT-based water supply monitoring and controlling system. The authors made use of a raspberry pi controller, relay module, water flow sensor and solenoid valve. The programming language used was python language to input commands into the raspberry pi controller to monitor the water supply. Narendran *et al.* [30] proposed an internet of things-based sustainable water management system to automate the water management process through real-time optimization of water using the real-time data collected from the field. In the development of an affordable IoT-based water management system for a large campus, Verma *et al.* [31] focused on a low-cost ultrasonic-based water-level sensor and a sub-GHz-based campus scale, wireless network to connect the sensors. The wireless network made use of sub-GHz radios to connect to a gateway to upload the water level data to the cloud for visualization and analytics. They concluded by suggesting the inclusion of more control valves to create a smart distribution system for the distribution of enough water. In another research by Eisha [32], the modeling of a smart water control mechanism using IoT was proposed. The model was implemented using an analytical approach with cost-effective and functional modules, which made use of sensors and wireless communication systems. This mechanism was designed mainly for water distribution and monitoring applications. In a water demand forecast research carried out by Lakshmi and Suresh [33], the authors proposed an IoT-based water demand forecasting and distribution design for a smart city by carrying out a three-month daily observation of water demand using ARIMA and regression to analyze the case study. Based on the water demand forecasting analysis, they carried out the IoT-based architecture using hydraulic engineering design for the proper distribution of water with minimal losses. This resulted in the development of a smart water distribution system (SWDS). In a study by Siddula *et al.* [34] on water level monitoring and management of dams using IoT, the authors proposed a system that could provide water levels in real-time and that could give conclusions on the safe operations of dams. They proposed the idea of collecting and sharing real-time information about water levels to an authorized central command centre using Bluetooth communication modules. The authorized centre command would determine the opening or closing of the dam gates. The proposed system automates the control of the dams without human interference using an ultrasonic sensor interfaced with a microcontroller. To meet water requirements, distribution and quality checks, Joy [35] proposed an IoT approach consisting of different sensors like the water flow sensor, pH sensor, temperature, water control valve and raspberry PI controller. The water flow sensor controls the water control valve to ensure equal and adequate water distribution to each end. Geetha *et al.* [36] developed a real-time water quality monitoring system, using the internet of things to test water samples and upload results over the internet. Whenever there is a deviation in water quality parameters from the pre-defined set values, the user will be alerted. The model aimed to present a low-cost and less complex smart water quality monitoring system using a controller with an inbuilt Wi-Fi module to monitor parameters such as pH, turbidity and conductivity. Kawarkhe and Sanjay [37] proposed a smart water monitoring system for domestic use to measure water pH, water level, flow, temperature and other parameters. The authors proposed a smart sensor interface device capable of monitoring water level, water pollution and water pipeline leakage. They made use of an ultrasonic sensor to check the water tank level and a temperature sensor to check the temperature of the water. They also made use of LabVIEW software for the automation feature, where laptops and mobile phones were used to control the automated system. The authors placed the system in a smart building and were able to collect and analyze the water patterns of the residents thereby preventing water waste. Apart from water level, flow, temperature and pH level, parameters like alternating voltage, current and temperature can also be monitored to aid water management. Uma *et al.* [38] developed a system capable of monitoring these parameters to control and monitor a submersible motor and upload the parameters to a central server. The system aimed to prevent the motor from burning out during a short circuit condition. Khaled *et al.* [39] introduced the notion of water level monitoring and management within the context of the electrical conductivity of the water. The authors investigated microcontroller-based water level sensing and controlling in a wired and wireless environment. They proposed a web-and cellular-based monitoring service protocol to determine and sense water levels globally. The system integrates the GSM module to alert the user. Oghogho and Azubuike [40] developed an electric water pump controller and level indicator using five metallic contact probes as level sensors. They connected the lowest probe to a 5 V power source to provide a fixed reference voltage. The other four probes served as inverting inputs into the comparators (Analogue to Digital Converter). By utilizing the conductivity of water when ionized, the probes were able to monitor the water level. Praseed *et al.* [41] made use of a Proportional-Integral-Derivative (PID) controller based on LabVIEW and Matlab software to obtain a liquid-level controller. The process of water pumping in overhead tank storage was automated taking advantage of the electrical conductivity property of water. The logic gate-based automatic water-level controller designed by Abrar and Rajendra [42] was based on an electro-mechanical system using digital technology. Electrical probes were inserted inside the tank. The probes detected the level of water and turned the motor ON/OFF as required. In a paper by Okhaifoh *et al.* [43], a controller-based automatic control for the water pumping machine and level indicator was designed, constructed and tested. The system used the reflection of sound (echo) to indicate of the water level in a storage tank. The MBACWPMLI used an ultrasonic sensor installed at the top of a tank to send and receive sound waves, and the time taken was converted to distance by the microcontroller to give corresponding digital outputs, which showed on an LED indicating the level of water in the storage tank. The IoT-based concept has also been applied in the oil sector by Rojiha [44] who proposed a sensor network-based automatic control system for efficient oil pumping unit management and well health monitoring. The author made use of several basic sensors for good data sensing while a microcontroller processed the data and used it to control the oil-pumping unit accordingly. The system was designed in such a way that at the detection of any abnormality an error SMS would be sent *via* GSM to the manager. The author concluded by saying that the system was efficient in monitoring and controlling oil wells remotely.

In the previous studies, the researchers were focused on the water resources being monitored therefore neglecting the pumping or supply system. The pumping systems themselves play a vital role in ensuring proper water distribution and management. When a pump is faulty, it is almost impossible to ensure the availability of water to consumers. For example, when a submersible pump gets faulty, it is usually difficult to trace the faults or detect the faults on time because it is submerged. This study is focused on monitoring the submersible pump to detect any abnormalities in real time to ensure a quick response, diagnosis and repair of faults. The succeeding sections present the materials and method, results and discussion of the findings as well as recommendations and conclusion.

MATERIALS AND METHOD

The isometric view of the water distribution and management system is shown in Fig. (**1**).

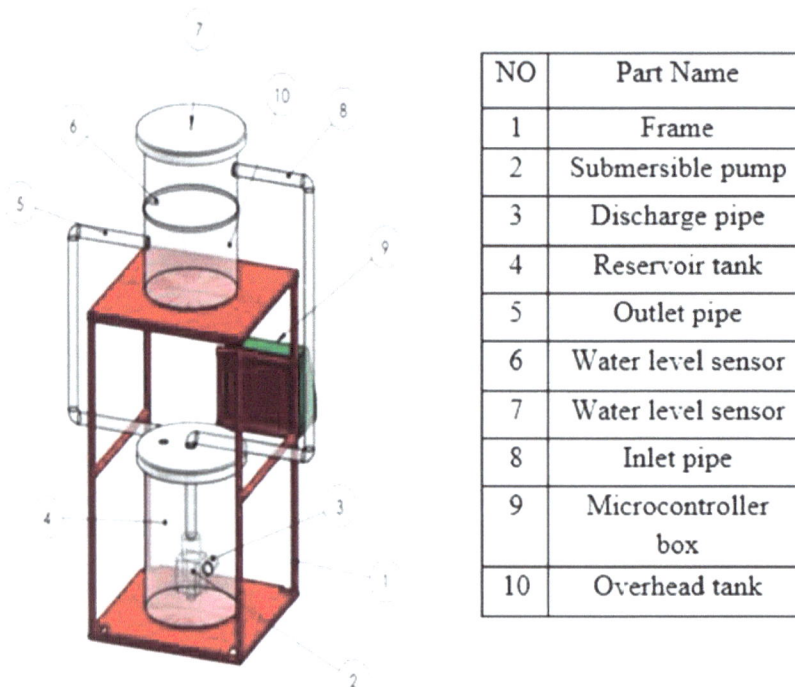

NO	Part Name
1	Frame
2	Submersible pump
3	Discharge pipe
4	Reservoir tank
5	Outlet pipe
6	Water level sensor
7	Water level sensor
8	Inlet pipe
9	Microcontroller box
10	Overhead tank

Fig. (1). Isometric view of the water distribution and management system.

Materials

The materials used for the development of the automatic water level monitoring system prototype are presented in Table **1**.

Table 1. Materials employed for the development of the automatic water level monitoring system.

S/N	Material	Description	Quantity
1.	Submersible pump	12V DC	1
2.	Arduino Nano board Microcontroller	ATmega328P, 8-bit, Input range: 7-12V	1
3.	Relay Module	-	1
4.	Liquid Crystal Display	12C 1602	1
5.	Piezoelectric Buzzer	-	-
6.	Lithium polymer battery (power supply)	12V, 10,500 mAH	-
7.	Arduino water level sensor	-	1
8.	Control Clip	-	-
9.	Water tanks and tubes	-	-
10	Electrical components (jumper wires, resistors, LEDs, breadboard).	-	-

Pump Design for Selection

A 12V DC submersible pump was used. The specifications of the pump, tank and pipe used for the system's development are presented in Table **2**.

The input current is expressed as Eq. (1).

$$I = \frac{P}{V} \tag{1}$$

Where: I is the current in Ampere, P is the power in watt and V is the voltage in volts.

For an input power of 4.8 W and voltage of 12 V, the input current is calculated as 400 A from Eq. (1).

The volume of the tank (v) is given in the equation as Eq. (2).

$$v = \pi r^2 h \tag{2}$$

Where: r is the tank radius (m), h is the height of the tank (m), $\pi = \frac{22}{7}$.

Given that the radius of the tank is 0.13 m at a height of 0.26 m, then the volume of the tank is calculated as 0.0138 m^3.

Considering the fact that the volumetric flow rate (Q) of the pump is 6.66 x $10^{-5}m^3$ / s (Table **2**), the pressure in the tank (P_t) is expressed as Eq. (3).

Table 2. The specifications of the pump, tank and pipe employed.

S/N	Parameter	Specification
-	**Pump**	-
1.	Input power (V)	4.8
2.	Input voltage (V)	12
3.	Maximum head	3.0
4.	Volumetric flow rate (m³/s)	6.66×10-5
5.	Inlet/outlet diameter (mm)	8
	Tank	-
1.	Height (m)	0.26
2.	Diameter (m)	0.26
3.	Radius (m)	0.13
	Pipe	-
1.	Internal diameter (mm)	6.35
2.	Outer diameter (mm)	8

$$P_t = P_0 + \rho g H \tag{3}$$

Where P_0 is the atmospheric pressure which is zero for a closed system, ρ is the density of water (1000 kg/m³) and g is the acceleration due to gravity (9.81 m/s²).

The height of the tank is 0.26 m, hence, the pressure in the tank (P_t) is calculated as 2550.6 N/m² from Eq. (3).

Eq. (4) expresses the force in the tank.

$$F = P_t \times A \tag{4}$$

Where A is the tank cross sectional area expressed as Eq. (5).

$$A = \pi r^2 \tag{5}$$

For a tank radius of 0.13 m, the tank cross sectional area is calculated as 0.053 m² from Eq. (5). Hence for a tank pressure of 2550.6 N/m², and cross-sectional area of 0.053 m², the force in the tank is calculated as 135.18 N from Eq. (4).

Eq. (6) expresses the efficiency of the pump.

$$\eta = \frac{\rho g Q H}{P} \times 100\% \tag{6}$$

For a flow rate Q (6.66×10-5 m³/s), acceleration due to gravity g (9.81 m/s²), maximum head H (3 m), power input, P (4.8 W), the efficiency of the pump is calculated as 40.9% from Eq. (6).

The time taken (t) to fill the empty overhead tank is expressed as Eq. (7).

$$t = \frac{v}{Q} \tag{7}$$

Recall the volume of the tank is 0.0138 m³ from Eq. (2) and volumetric flow rate Q is 6.66×10-5 m3/s. Thus, the time taken (t) to fill the empty overhead tank is calculated as 206.98 sec from Eq. (7).

Global System for Mobile Telecommunication (GSM) Module

GSM is an open, digital cellular technology used for transmitting mobile voice and data services. It is a mobile communication modem. It operates at 850 MHz, 900 MHz, 1800 MHz and 1900 MHz frequency bands. A GSM digitizes and reduces the data, then sends it down through a channel with two different streams of client data, each in its particular time slot. For this project, the SIM900A module was used. The specifications of the SIM900A module are given in Table **3**.

Table 3. Specification of the SIM900A module.

S/N	Item	Remarks
1	Dual-Band (MHz)	900/1800
2	GPRS multi-slot	Class 10/8
3	GPRS mobile station	Class B
4	Dimensions (mm)	24 x 24 x 3
5	Weight (g)	3.4
6	Control	AT commands (GSM 07.07, 07.05)
7	Supply voltage range (V)	3.2- 4.8
8	Low power consumption (mA)	1.0 (sleep mode)
9	Operation temperature (˚C)	-400 to +85

Arduino Nano Microcontroller

It is a small, complete and breadboard friendly board based on the ATmega328P. It provides a simple and modular way of interfacing real world with the computer to handle basic processing tasks on a chip while working with hardware sensors. The technical specifications of the Arduino Nano used for this study is given in Table **4**.

Table 4. Technical specification of the Arduino nano microcontroller.

S/N	Item	Remarks
1	Microcontroller processor	ATmega328P
2	Operating voltage (V)	5
3	Input voltage (recommended) (V)	7-12
4	Input voltage (limits) (V)	6-20
5	Digital I/O pins	14 for PWM output
6	Analog input pins	8
7	Power consumption (mA)	19
8	DC per I/O pin (mA)	40
9	DC for 3.3V pin (mA)	50
10	Flash memory (KB)	32 (2 used by bootloader)
11	SRAM/ EEPROM (KB)	2 /1
12	Clock speed (MHz)	16

Relay Module

A relay is an electrically operated switch that can be switched ON or OFF, to control the flow of current, and can be controlled with voltages as low as 5 v provided by the Arduino pins. A relay is operated by an electromagnet. The electromagnet requires a small voltage to activate and once it is activated, it pulls the contact to make the high-voltage circuit. The Arduino relay module has six pins: three on one side and three on the other side. On the bottom side, there are three pins, signal, 5 v and ground. There are NC (Normally closed), C (Common) and NO (normally open) which are the output pins of the 5 V relay. The Arduino relay module can be used in two states; Normally Open State (NO) and Normally Closed State (NC).

In the Normally Open (NO) state, the initial output of the relay will be low when it is powered. In this state, the common and the normally open pins are used. The normally open configuration works in such a way that the relay is always open.

This means that the circuit is broken unless a signal is sent from the Arduino to close the circuit. For this study, the relay module is in NO state.

In the Normally Closed (NC) state, the initial output of the relay will be high when it is powered. In this state, the common and the normally close pins are used. COM: common pin. The normally closed configuration is used when there is the need for the relay to be closed by default. This means that the current continues to flow until a signal is sent from the Arduino to the relay module in order to open the circuit and stop the current flow.

Piezo Buzzer

Piezo buzzer is an electronic device commonly used to produce sound. The buzzer makes use of a piezoelectric material. When subjected to an alternating electric field it stretches or compresses in proportion to the frequency of the signal thereby producing sound.

Arduino Water Level Sensor

It was designed for water detection, which can be widely used in sensing rainfall, water level, and even liquid leakage. It is easy to install and has minimal installation cost and maintenance. In addition, it can be installed in all tank types because it does not rely on the tank walls for its measurement function. A non-contact sensor like the ultrasonic proximity sensor was not used because of the size and volume of the tank because the signal could bounce off the tank's walls and give false readings. The technical specifications of the Arduino water level sensor are shown in Table **5**.

Table 5. Technical Specification of the Arduino water level sensor.

S/N	Item	Remarks
1	Working voltage (V)	5
2	Working current (mA)	<20
3	Interface	Analog
4	Width of detection (mm)	40×16
5	Working Temperature (°C)	10~30
6	Weight (g)	3
7	Size (mm)	65×20×8
8	Arduino compatible interface	Yes
9	Low power consumption	Yes

(Table 5) cont.....

S/N	Item	Remarks
10	High sensitivity	Yes
11	Output voltage signal (V)	0~4.2

Liquid Crystal Display (LCD)

In this study, a 12C 1602 LCD was used to display activities in the motherboard.

Method

In this study, the relay was used in the normally open configuration. The relay switches ON when a LOW signal is sent from the Arduino but switches OFF when a HIGH signal is sent. The relay works with an inverted logic. The relay module was programmed in a way that it switches ON/OFF the pump at the predetermined water level. It serves as a switch as it enables the pump to start on its own. Three Arduino water level sensors were required to detect the water level.

1. The first sensor was placed in the pipe transporting water from the reservoir tank. The sensor detects the flow of water from the reservoir to the overhead tank. The Arduino uses the signal from this sensor to detect the functionality of the pump.

2. The second sensor was placed at the predetermined minimum water level in the water tank at which the pump is expected to be switched on automatically when water level goes down below such presetting. The Arduino makes use of the signal from this sensor to detect when to switch ON the pump using the relay switch.

3. The third sensor was placed at the top of the overhead tank. This sensor detects the maximum water level in the tank. The Arduino makes use of this signal to switch OFF the pump.

The LCD has a parallel interface; this means that the microcontroller has to manipulate several interface pins at once to control the display. The LCD was programmed to display the following pump's status at the predetermined levels;

1. The LCDs display "Full, the pump stops", when the water level gets to the maximum level.

2. The LCD displays "Half, pump starts", when the water gets to the middle of the tank.

In the context of this study, the buzzer serves as an alarm to draw the attention of the user close to the system. The frequency of the sound was programmed such that: at low water level before the water gets to the high level, the buzzer produces a continuous beeping sound at an interval of 500 μs. When it gets to the high-water level, the beep changes tune and frequency at an interval of 1000 μs but when an error occurs, the buzzer makes a completely different sound. The circuit connection of the Arduino, water level sensors, piezo buzzer and LCD is as shown in Fig. (**2**).

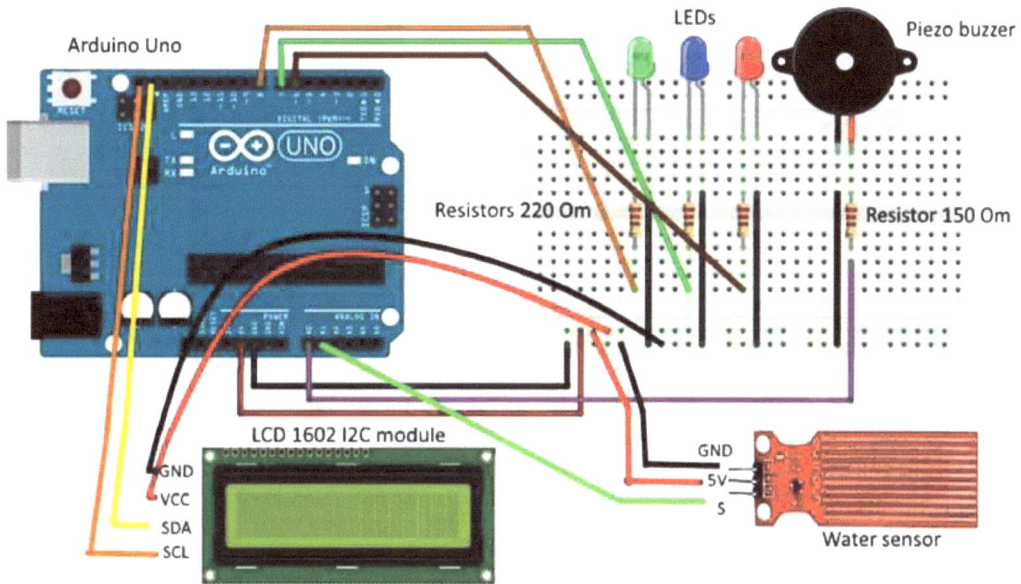

Fig. (2). Complete circuit connection.

The following were the steps taken to interface the GSM module with the Arduino: First, the band rate of the GSM Module was initialized; then, the first half of a serial Print "\r" starts the message and is followed by a second delay every time an attention (AT) command is being used; the GSM Module was set to Text Mode; the number to send the SMS to was inputted and the message "The pump is not functioning well. Please check the pump" was inputted.

RESULTS AND DISCUSSION

Testing and analysis of the system were done to ensure a fully functional study with minimum flaws as possible. The system was tested by pumping water into the over-head tank from the reservoir tank using laboratory demonstration of the component shown in figure 1. The water was moved from the reservoir tank with the use of a submersible pump. The setup represented an actual borehole submersible pump supplying water to a domestic over-head tank. The results obtained are presented in Tables **6** and **7**.

Table 6. Table for the values of volume and time taken to fill the tank at constant flow-rate.

S/N	Radius of tank (m)	Height of tank (m)	Volume of tank (m³)	Time taken to fill empty tank (sec)
0	0.13	0.26	0.01	206.95
1	0.16	0.29	0.02	349.65
2	0.19	0.32	0.04	544.07
3	0.22	0.35	0.05	797.83
4	0.25	0.38	0.07	1118.57
5	0.28	0.41	0.10	1513.91
6	0.31	0.44	0.13	1991.48
7	0.34	0.47	0.17	2558.91
8	0.37	0.5	0.21	3223.83
9	0.4	0.53	0.27	3993.88
10	0.43	0.56	0.33	4876.68
11	0.46	0.59	0.39	5879.86
12	0.49	0.62	0.47	7011.05
13	0.52	0.65	0.55	8277.88
14	0.55	0.68	0.65	9687.99
15	0.58	0.71	0.75	11248.99
16	0.61	0.74	0.86	12968.52
17	0.64	0.77	0.99	14854.22
18	0.67	0.8	1.13	16913.71
19	0.7	0.83	1.28	19154.61
20	0.73	0.86	1.44	21584.57

NB (The flow rate Q is constant at $6.66 \times 10^{-5} m^3/s$).

Table 7. The water tank level measurement using Arduino measuring sensor.

S/N	Water tank level (%)	LCD Indicator	Pump Operating Mode	Pump Operating Condition	GSM Module
1	100	Pump stop	OFF	OK	-
2	90	-	ON	-	-
3	80	-	ON	-	-
4	70	-	ON	-	-

(Table 7) cont.....

S/N	Water tank level (%)	LCD Indicator	Pump Operating Mode	Pump Operating Condition	GSM Module
5	60	-	ON	-	-
6	50	Pump on	ON	OK	-
7	40	-	OFF	NOT OK	Please check the pump
8	30	-	-	-	-
9	20	-	-	-	-
10	10	-	-	-	-
10	0	-	-	-	-

Fig. (**3**) presents the volume and time taken to fill the tank at constant flow rate of $6.66 \times 10^{-5} m^3/s$. The results obtained indicate a linear but inverse relationship between the volume of tank and the time taken to fill the tank. It is necessary to regulate the pump with the aid of the microcontroller once the water volume falls below the threshold, so that time will not be wasted in filling the tank.

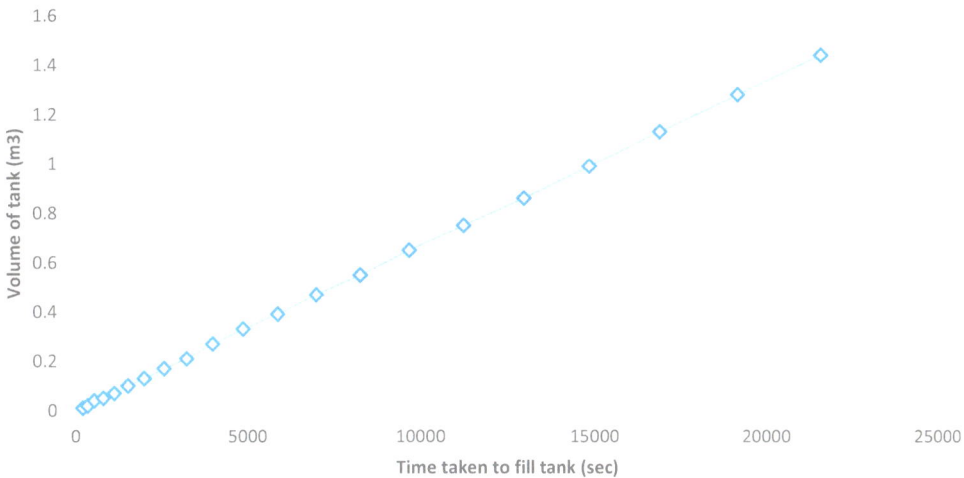

Fig. (3). Volume and time taken to fill the tank at constant flowrate.

The water tank level was measured using a conductive level measuring sensor "the Arduino water level sensor". This sensor was chosen for this study because of its ability to eliminate false or inaccurate readings. From Table **7**, the three Arduino water level sensors were tested and gave good results at 100% to 10% water volumes, the pump was switched OFF to preserve the energy consumption and water overflow at 100% WL. The sensor was able to measure accurately the level of water in the tank and displayed the same on the LCD screen. The microcontroller was able to regulate the switching OFF and ON of the pump at 100% and 50%, WL respectively based on the signals from the water level sensors and relay. In addition, the Arduino microcontroller was able to alert the user *via* the GSM module when the pump was not functioning properly, that is when the volume of the water is below 50%, and the pump is not in the switch ON mode.

The GSM module serves as a feedback mechanism, alerting the user of the pump's availability. The programming code for the module was written, simulated and ran. A successful code was achieved by detecting and correcting all syntax errors. The efficiency of the GSM module was tested, and it was able to produce good results. It was able to detect and alert the user *via* a text when the pump was not sending water to the overhead tank. This proved that the maintenance team would be alerted as soon as the pumping system develops a fault and would ensure quick response and repair or replacement before the water in the overhead tank would be empty. This will help the user and maintenance team to plan sufficiently to repair or replace the pump in case of failure and by so doing achieve 100% pump availability hour. Previous studies focused on using smart technology to manage water distribution systems by detecting water leakage, eliminating water tank overflow and monitoring pump energy consumption [22, 26, 43, 45]. This study is novel as it helps the user and the maintenance team to detect in real time when the pump fails through the GSM module. Although, when the module was tested in certain areas; it was not able to perform due to its high-frequency band of 900/1800 MHz. According to a report by Henry [46], some network service providers are yet to increase their operating frequency bands to 900/1800 MHz in some places in Nigeria.

CONCLUSION

This study has been proven to promote the availability of water distribution by using smart technology to detect the problem of the faulty pump in real-time. The performance evaluation conducted by the study revealed that when the submersible pump failed to lift water to the tank at less than 50% WL tank alert is sent to the user/maintenance team to check the pump. This will enable the problem associated with the pump to be repaired or replaced before the water in the tank is emptied. The system is user-friendly and environmentally friendly. The microcontroller was able to regulate the switch OFF and ON the pump at 100% and 50%, WL respectively based on the signals from the water level sensors and relay. In addition, the Arduino microcontroller was able to alert the user *via* the GSM module when the pump was not functioning properly, that is when the volume of the water is below 50%, and the pump is not in the switch ON mode.

Future studies can consider the development of a system with the capability to operate in different modes namely: manual, semi-automatic and fully automatic modes.

REFERENCES

[1] "UNDP. Human development report, sustaining human progress: Reducing vulnerability and building resilience", *United Nations Development Programme,* 2014. Available at: http://hdr.undp.org/en/content/human-development-report-2014 (Accessed on: 5 May 2020).

[2] "UN office of the high commissioner for human rights", *The Right to Water,* 2010. Available at: https://www.Refworld.Org/Docid/4ca45fed2.Html (Accessed on: 26 February 2020).

[3] A.J. Idu, "Threats to water resources development in nigeria", *J Geol & Geophys,* vol. 4, no. 3, pp. 1-10, 2015.
[http://dx.doi.org/10.4172/2381-8719.1000205]

[4] V.E. Ejiofor, and F.O. Oladipo, "Microcontroller based automatic water level control system", *IJIRCCE,* vol. 1, no. 6, pp. 1390-1396, 2013.

[5] WHO world health organization, *World Water Day Report,* 2015. Available at: http://Www.Who.Int/Water_Sanitation_Health/Takingcharge.Html (Accessed on: 2 February 2020).

[6] "WHO and UNICEF. 1 In 3 People Globally Do Not Have Access to Safe Drinking Water", Available at: https://Www.Who.Int/News-Room/Detail/18-06-2019-1-In-3-People-Globally-Do-Not-Have-Access-To-Safe-Drinking-Water-Unicef-Who (Accessed on: 14 December 2019).

[7] H.T. Ishaku, M.R. Majid, A.A. Ajayi, and A. Haruna, "Water supply dilemma in nigerian rural communities: Looking towards the sky for an answer", *J. Water Resource Prot.,* vol. 3, no. 8, pp. 598-606, 2011.
[http://dx.doi.org/10.4236/jwarp.2011.38069]

[8] "UN united nations", *The Right to Water,* 2010. Available at: https://www.Un.Org/En/Sections/Issues-Depth/Water/ (Accessed on: 10, February 2020).

[9] "UNDP. Human development report", In: *Millennium Development Goals: A Compact among Nations to End Human Poverty.* Oxford University Press.: New York, NY, 2003, pp. 1-384.

[10] D.I-L. Kelechi, *"The relationship between water scarcity awareness by transnational nigerians and its effect on the willingness to contribute to water scarcity relief efforts in nigeria"*, MSc Thesis in Community, University of Illinois, US, 2012.

[11] "Food and agriculture organization. hot issues: water scarcity", Available at: http://www.Fao.Org/Nr/Water/Issues/Scarcity.Html (Accessed on: December 7, 2019).

[12] "UNICEF nigeria water and sanitation profile. Water, Sanitation and Hygiene", Available at: https://Www.Unicef.Org/Nigeria/Water-Sanitation-And-Hygiene (Accessed on: 7 December 2019).

[13] O.A. Nelson, "How nigeria is wasting its rich water resources", Available at: https://Theconversation.Com/How-Nigeria-Is-Wasting-Its-Rich-Water-Resources-83110 (Accessed on: December 7, 2019).

[14] K.N. Pradeep, "Water crisis in africa: myth or reality international journal of water resources development", *Int. J. Water Resour. Dev.,* vol. 33, no. 2, pp. 326-339, 2016.

[15] R.M. Olalekan, O. Adedoyin, A. Ayibatonbira, B. Anu, O.O. Emmanuel, and N.D. Sanchez, "Digging deeper evidence on water crisis and its solution in nigeria for bayelsa state: A study of current scenario", *Int. J. Hyd.,* vol. 3, no. 4, pp. 244-257, 2019.
[http://dx.doi.org/10.15406/ijh.2019.03.00187]

[16] Y. Tu, R. Wang, Y. Zhang, and J. Wang, "Progress and expectation of atmospheric water harvesting", *Joule,* vol. 2, no. 8, pp. 1452-1475, 2018.

[http://dx.doi.org/10.1016/j.joule.2018.07.015]

[17] A.N. Jerry, "Water supply system", *Encyclopedia Britannica,* 2020. Available at: https://www.britannica.com/technology/water-supply-system (Accessed on: May 12, 2020).

[18] M.K. Faizah, "Water supply and distribution", Available at: https://www.academia.edu/27278221/WATER_SUPPLY_AND_DISTRIBUTION (Accessed on: December 7, 2019).

[19] "National research council. Flouride action network", Available at: https://fluoridealert.org/researchers/nrc/ (Accessed on: 28th May, 2022).

[20] S. Pallavi, and R.S. Smruti, "Internet of things: Architectures, protocols, and applications", *J. Electr. Comput. Eng.,* pp. 1-26, 2017.

[21] R. Varsha, and W. Wu, "IoT technology for smart water system", *IEEE 20th International Conference on High Performance Computing and Communications; IEEE 16th International Conference on Smart City; IEEE 4th Intl. Conference on Data Science and Systems* Exeter, UK pp. 1493-1498, 2018.

[22] M. Odiagbe, E.M. Eronu, and F.E. Shaibu, "An effective water management framework based on internet of things (IOT) technology", *Eur. J. Eng. Sci. Tech.,* vol. 4, no. 5, pp. 102-108, 2019. [http://dx.doi.org/10.24018/ejers.2019.4.5.1317]

[23] M. Nunes, R. Alves, A. Casaca, P. Póvoa, and J. Botelho, "An internet of things based platform for real-time management of energy consumption in water resource recovery facilities", In: *Internet of Things. Information Processing in an Increasingly Connected World. IFIPIoT 2018. IFIP Advances in Information and Communication Technology,* L. Strous, V. Cerf, Eds., vol. 548. Springer, Cham., 2018.

[24] I.A. Daniyan, O.L. Daniyan, K. Mpofu, and B.I. Ramatsetse, (2019). Development and Performance Evaluation of Automated Irrigation System. Published Proceedings of the SAUPEC/RobMech/PRASA Conference Bloemfontein, South Africa, January 28-30, 29. Added to IEEE Xplore, 978-1-7281-0-69-3/19. pp. 12-16.

[25] L. Daniyan, E. Nwachukwu, I. Daniyan, and O. Bonaventure, "Development and optimization of an automated irrigation system", *J. Autom. Mob. Robot. Intell. Syst.,* vol. 13, no. 1, pp. 37-45, 2019. [http://dx.doi.org/10.14313/JAMRIS_1-2019/5]

[26] M. Hadipour, J.F. Derakhshandeh, and M.A. Shiran, "An experimental setup of multi-intelligent control system (MICS) of water management using the internet of things (IoT)", *ISA Trans.,* vol. 96, pp. 309-326, 2020. [http://dx.doi.org/10.1016/j.isatra.2019.06.026] [PMID: 31285060]

[27] C. Kamienski, J.P. Soininen, M. Taumberger, R. Dantas, A. Toscano, T. Salmon Cinotti, R. Filev Maia, and A. Torre Neto, "Smart water management platform: iot-based precision irrigation for agriculture", *Sensors,* vol. 19, no. 2, p. 276, 2019. [http://dx.doi.org/10.3390/s19020276] [PMID: 30641960]

[28] T. Cristina, T. Cornel, and G. Vasile, "An internet of things oriented approach for water utility monitoring and control", *Adv. Comp. Sci.,* pp. 175-180, 2018.

[29] H.V. Maruthi, A.R. Lavanya, M. Meda, and P.L.J. Lakshmi, "An IoT-based water supply monitoring and controlling system", *Int. J. Adv. Res. Comput. Sci.,* vol. 9, no. 3, pp. 202-206, 2018.

[30] S. Narendran, P. Pradeep, and M.V. Ramesh, "An internet of things (IoT) based sustainable water management", *2017 IEEE Global Humanitarian Technology Conference (GHTC)* San Jose, CA pp. 1-6, 2017. [http://dx.doi.org/10.1109/GHTC.2017.8239320]

[31] P. Verma, K. Akshay, R. Nihesh, J. Pratik, S. Mallikarjun, R. Subramanian, B. Amrutur, M.S.M. Kumar, and R. Sundaresan, "Towards an IoT based water management system for a campus", *2015 IEEE First International Smart Cities Conference (ISC2)* Guadalajara, Mexico 2015, pp. 1-6. [http://dx.doi.org/10.1109/ISC2.2015.7366152]

[32]   D.E. Akanksha, "Modeling of Smart Water Control Mechanism using IoT", *Int. J. Innov. Technol. Explor. Eng.,* vol. 9, no. 2, pp. 149-154, 2019.
[http://dx.doi.org/10.35940/ijitee.A5266.129219]

[33]   K.N. Lakshmi, and S. Suresh, "IoT-Based Water Demand Forecasting and Distribution Design for Smart City", *J. Water Clim. Chang.,* vol. 11, no. 4, pp. 1411-1428, 2019.

[34]   S.S. Siddula, P. Babu, and P.C. Jain, "Water level monitoring and management of dams using IoT", *3rd International Conference on Internet of Things: Smart Innovation and Usages (IoT-SIU),* 2018 pp. 1-5
[http://dx.doi.org/10.1109/IoT-SIU.2018.8519843]

[35]   S. Joy, "An internet of things based model for smart water distribution with quality monitoring", *Int. J. Innov. Res. Sci. Eng. Technol.,* vol. 6, no. 3, pp. 3446-3451, 2017.

[36]   S. Geetha, and S. Gouthami, "Internet of things enabled real time water quality monitoring system", *Smart Water 2,* 2016. Available at: https://Doi.Org/10.1186/S40713-017-0005-Y (Accessed on: February 4, 2020).

[37]   M.B. Kawarkhe, and A. Sanjay, "Smart water monitoring system using IOT at home", *IOSR J. Comput. Eng.,* vol. 21, no. 1, pp. 14-19, 2019.

[38]   R.K. Uma, K. Prashanth, K. Vijay, and S. S. Begam, "IoT based project for submersible motor controlling, monitoring, & updating parameters to central server with free rtos", *International Research Journal of Engineering and Technology,* vol. 4, no. 6, pp. 2083-2085, 2017.

[39]   R. Khaled, S. Ahsanuzzaman, M. Tariq, and R.S.M. Mohsin, "Microcontroller based automated water level sensing and controlling: Design and implementation issue", *Proceedings of the World Congress on Engineering and Computer Science* US 1:1-5, 2010.

[40]   I. Oghogho, and C. Azubuike, "Development of an electric water pump controller and level indicator kwara state", *Int. J. Eng. Appl. Sci.,* vol. 3, no. 2, pp. 18-21, 2013.

[41]   K. Praseed, S.S. Pathan, and B. Mashilka, "Liquid level control using pid controller based on labview & matlab software", *Int. J. Eng. Res. Technol.,* vol. 3, no. 10, pp. 111-114, 2014.

[42]   M.M. Abrar, and R.P. Rajendra, "Logic gate based automatic water level controller", *Int. J. Res. Eng. Technol.,* vol. 3, no. 4, pp. 477-482, 2014.
[http://dx.doi.org/10.15623/ijret.2014.0304085]

[43]   J. E. Okhaifoh, C. K. Igbinoba, and K. O. Eriaganoma, "Microcontroller based automatic control for water pumping machine with water level indicators using ultrasonic sensor", *Niger. J. Technol.,* vol. 35, no. 5, pp. 579-583, 2016.
[http://dx.doi.org/10.4314/njt.v35i3.16]

[44]   C. Rojiha, "Sensor network based automatic control system for oil pumping unit management", *Int. J. Sci. Res. Publ.,* vol. 3, no. 3, pp. 1-4, 2013.

[45]   Radhakrishnan, V. & Wu, W. "IoT Technology for Smart Water System," 2018 IEEE 20th International Conference on High Performance Computing and Communications; IEEE 16th International Conference on Smart City; IEEE 4th International Conference on Data Science and Systems (HPCC/SmartCity/DSS), Exeter, United Kingdom, 2018, pp. 1491-1496.

[46]   "Henry Lancaster Nigeria Mobile Infrastructure, Operators and Broad bands-Statistics and Analyses", Available at: http://Www.Budde.com.au/Research/ Nigeria-Mobile-Infrastructure-Operators-and-Broad-band-Statistics-and-Analyses (Accessed on: July 1, 2020).

CHAPTER 10

Development of Automated Waste Segregator

Ilesanmi Afolabi Daniyan[1,*]**, Adefemi Adeodu**[2]**, Jacobs Kelechi**[3] **and Lanre Daniyan**[4]

[1] *Department of Industrial Engineering, Tshwane University of Technology, Pretoria 0001, South Africa*

[2] *Department of Mechanical Engineering, University of South Africa, Florida, South Africa*

[3] *Department of Mechanical & Mechatronic Engineering, Afe Babalola University of Nigeria, Ado Ekiti, Nigeria*

[4] *Department of Instrumentation, Centre for Basic Space Science, University of Nigeria, Nsukka, Nigeria*

Abstract: One of the major problems with waste generation today is the huge percentage of plastics in its composition. The automated waste segregator is a mechatronic system that solves this problem by the incorporation of high calibrated sensors and mechanical properties into its design to enable the smooth segregation of plastics and metals. It is capable of detecting and separating these components as soon as they get to its sensing unit with the aid of capacitive proximity sensors and ultrasonic sensors. The capacitive proximity sensor made with plated iron detects the type of material in range either plastics or metals based on their di-electric constant. A 12V DC geared motor made with aluminum and iron of high torque and speed characteristics is connected to the lids to enable its automatic open and close mechanism. Also, a microcontroller (PIC18F452) was used to control the entire segregation process of the system and is capable of storing and implementing the software in real-time. The sensors are controlled by the PICI8F452 microcontroller and based on the sensor readings, pulses are sent from the controller to the geared motor for the fast action of lid opening. The results of the performance evaluation indicated that the automated waste segregator has the capacity to identify and sort wastes into plastics, metals and any other waste with the aid of a capacitive sorting technique.

Keywords: Environmental Hazards, Microcontroller, Pollution, Proximity Sensors, Waste.

INTRODUCTION

The major problem with waste generation today is the huge percentage of plastics in its composition. The burning of the waste components produces a very toxic

---

* **Corresponding authors Ilesanmi Afolabi Daniyan:** Department of Industrial Engineering, Tshwane University of Technology, Pretoria 0001, South Africa; Tel: +27 (064) 5298778; E-mail: afolabiilesanmi@yahoo.com

gas known as dioxin that is dangerous to human health and can endanger green life as well as constitute environmental hazards. Many methods have been employed for carrying out waste segregation for various types of materials namely the manual methods, sensor implementation, electromagnetic, and x-ray sorting. The manual method involves hand-sorting which is harmful and laborious [1, 2]. On the other hand, the waste sorting based on the use of smart sensors utilizes the capacitive, inductive and infrared sensors to segregate waste depending on the characteristics and properties of wastes to be sorted [3]. An electromagnetic separator used for separating metals utilizes an eddy current when a conductor is exposed to magnetic field making it possible for metals to be separated from non-metals. The x-rays sorting method sorts the materials based on their atomic densities. It achieves a high-resolution level of sorting regardless of moisture and pollution level of waste to be sorted. The use of energy-resolved photon-counting detectors with multiple thresholds, which allow simultaneous measurements of the x-ray attenuation at multiple energies, allows separation [4-5]. The effective sorting out of waste materials will provide an opportunity for recycling and reuse of some waste products [6 - 8]. For instance, plastics and metals can be recycled and put to better use instead of causing environmental pollution but because of being mixed-up with other components in the waste stream, it makes their management extremely difficult. The need for proper waste management in the environment cannot be over-emphasized as waste management issues are interfering significantly with humans' overall well-being and society both on a large and small scale. Improper waste disposal can cause ailments like asthma, emphysema and other environmental hazards and pollution if not managed properly [9-11]. Waste is generally generated per seconds in the manufacturing industries, communities, schools and homes and hence the need for an efficient, effective and sustainable waste management system that incorporates proper sorting of waste components. The automated waste segregator is a mechatronic system that is able to separate waste namely; plastics, metals and house waste. Through the use of an analog sensor (capacitive proximity sensor), the system is able to detect plastic and metallic materials and separate them by the movement of electric motors and a conveyor system. The automated waste segregator is incorporated in an electronic bin making up a compact and effective system. Many researchers have reported on the development of automated waste sorting machines [12-15], the novelty of this work, however, lies in the compatibility of the system with Bluetooth and Android applications for querying and feedback. This will enhance real-time monitoring of the waste sorting activities.

MATERIALS AND METHOD

The materials employed for the purpose of this study include: a capacitive proximity sensor (with a sensing range of 2-40 mm for detecting both metal and non-metals), an ultrasonic sensor made of silicon-based PCB and partly aluminum, a 12V DC geared motor made with aluminum and iron of high torque and speed characteristics, a microcontroller (PIC18F452), a circuit board, a vero board, and L293D motor driver. The four sections employed in the design and fabrication of automated waste segregation are: electrical and electronic systems integration, fabrication and software implementation. The materials selection considers the service and functional requirement as well as the material's cost availability. The major components used in the electrical and electronics phase are the capacitive proximity sensor made with plated iron which detects the type of material in range either plastics and metals based on their dielectric constant; an ultrasonic sensor made of silicon-based PCB and partly aluminum, a 12V DC geared motor made with aluminum and iron of high torque and speed characteristics that is connected to the lids to enable its automatic open and close mechanisms, and a microcontroller (PIC18F452) used to control the entire segregation process of the system and capable of storing and implementing the software in real-time. This is the heart and brain of the system where all the peripheral devices are connected to the first analogue channel of the microcontroller (R1). The L293D pins are connected to the CCP pins of the microcontroller for speed control. The ultrasonic sensor is connected to RB7 and RB6 while the trigger is connected to RB7 and ECHO to RB. The GND pin is connected to the common ground and the Vee to +5V voltage source. The circuit board is used to hold all the electrical and microcontroller components of the system, connecting wires used to make connections of the circuitry on a Vero boar. An L293D motor driver is employed for controlling the motion of the electric DC motors. Other components in this phase include a dozen of 9 volts batteries, capacitors, resistors, a 5 V voltage regulator, a crystal oscillator (4 MHz), pushbuttons and integrated circuit chip sockets. Figs. (**1-3**) show the circuit design, its implementation and the block diagram of the system.

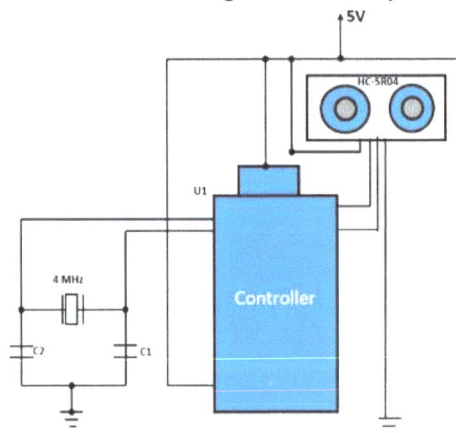

Fig. (1). Ultrasonic sensor interfacing with PIC18F452 microcontroller.

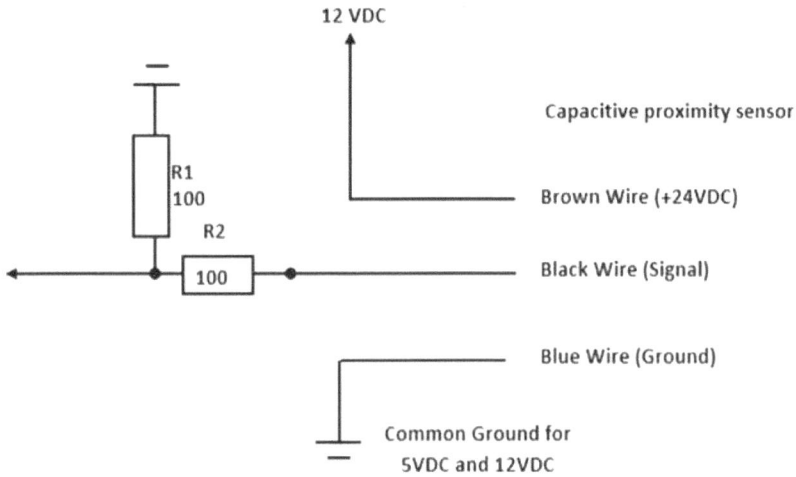

Fig. (2). Capacitive sensor interfacing with a PIC18F452 microcontroller.

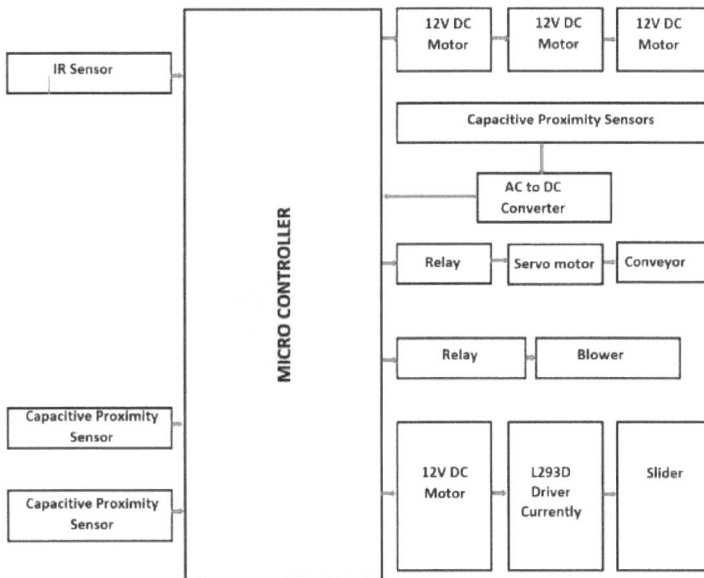

Fig. (3). Block diagram of the system.

The Assembly Phase

Aluminium is used for the main body frame and solid works are used for the design package because of excellent mechanical properties of high strength to weight ratio as well as wear and corrosion resistance. Punching, hammering, and joining (temporary) operations are performed on the prototype during fabrication.

The Software Phase

The C language is used for this study with the CCS C compiler version 5.0.0.1. The phase is divided into three steps: motor control, analog sensor reading and ultra-sensor distance measurement. Three DC motors are controlled in the program and this has been achieved by the use of the L293D H-bridge driver and the voltage reading from the analog sensor while the measure of the object's distance is made*via* the use of the ultrasonic sensor. There is one compartment that receives the input waste with three other compartments into which the wastes namely plastic, metallic and other wastes are stored. The capacitive sensor is a non-contact sensor that senses the type of material by noting the change in the capacitance read by the sensor. It has two conducting elements connected to the oscillator circuit and the output amplifier. The air gap between the two conducting elements represents the dielectric material (insulator). As the waste to be segregated reaches the sensing area, the capacitance of the element increases causing amplitude changes in the oscillator thus, creating an output signal that is fed into the microcontroller. The microcontroller in turn activates the DC motor. The DC motor drives the sorted input wastes toward their respective compartments. Based on this, the various lids will open at various voltage readings from the capacitive proximity sensor and ultrasonic sensor state. The motors remain in the OFF state if the ultrasonic sensor does not give a distance reading of 50 cm but as soon as the distance is attained, the microcontroller reads the voltage given by the capacitive proximity sensor. If a particular voltage is read by the microcontroller, based on the program, the microcontroller opens a particular lid, either the plastic, metals or other waste lid.

RESULTS AND DISCUSSION

The highest sensitivity level was set and it successfully passed all the tests in this phase. The compactness and detachability of the system test are major design objectives and it was successfully implemented. It is user friendly and anyone can work on the system even with little knowledge of the operation. The waste segregator which can house a large capacity of waste materials due to a large waste compartment was designed and constructed. The system which was compactly designed was used to separate waste into three sections namely: plastic, metallic and any other domestic waste. Based on the design of the system's circuitry and overall system design, the amount of components utilized for adequate operation of the system is minimally reduced. The main segregator unit can be detached and used on any other waste bin of equal size and shape. The bin is a fast-acting microcontroller-based system, very easy to use and understand. It is also capable of housing a reasonable amount of waste before disposal. Fig. (**4**) presents the developed automatic waste segregator.

Fig. (4). The developed automatic waste segregator.

The input waste fed into the segregator includes; paper, plastics, food waste, domestic wastes and metallic wastes amongst others. The categories of the wastes sorted according to their voltage range and their respective composition are presented in Table **1**.

Table 1. The categories of the wastes sorted and their percent composition.

Waste	Types of Wastes	% Composition
Iron scraps	Metallic	3
Zinc scraps	Metallic	2
Lead scraps	Metallic	4
Steel scraps	Metallic	5
Aluminum scraps	Metallic	6
Plastic bottles, jugs, cups and jars	Plastic	18
Compact discs	Plastics	2
Polythene bags, nylon	Plastics	9
Computer keyboards and frames	Plastics	2
Tooth brush	Plastic	2
Fibre and plastic tubing	Plastic	2
Yam, potato cassava and fruit peels	Others	10
Food and animal waste	Others	5
Leather wastes	Others	12
Papers	Others	14
Glasses and ceramics	Others	4

Fig. (**5**) shows the types of wastes and the percent composition. The Figure shows that a sizeable percentage of metallic waste (20%) and plastic waste (35%) have been sorted for recycling while the remaining 45% falls under the category of "other" types of waste. Hence, recycling will bring about sustainable means of waste management and a means to generate wealth from waste.

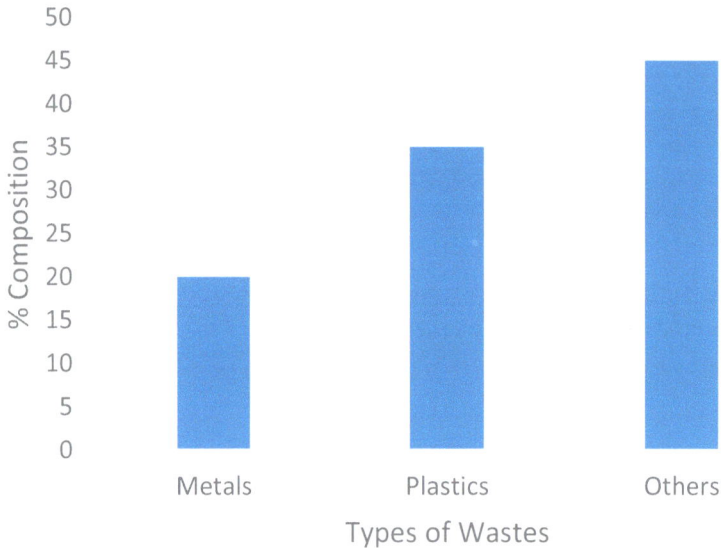

Fig. (5). Waste types and percent composition.

CONCLUSION

An automated waste segregator has been successfully tested and implemented for the segregation of waste into plastics, metallic and any other waste at a domestic level with the aid of a capacitive sorting technique. The study has proved it to be very reliable and it can be of great use in houses, offices, industrial settings and all environments where waste is been produced and effective waste management strategy is needed. It has the ability to segregate one component of waste at a time with an assigned priority for plastics and metallic wastes. The greatest challenge encountered in the design phase was the implementation of the ideology of a detachable segregation unit that could be used on any bin of equal size and because of this, the design was changed a number of times during the research.

The following points are recommended for further research considerations;

1. The system could be improved by enabling it to separate any waste by storage of numerous material capacitive values.

2. The system could be improved by adding a sensor and a buzzer to sound an alarm when the unit is filled with trash cans and rubbish.

3. A powerful crusher can be improved by added to the design to press/ compress plastic and aluminum cans either in the inlet section or inner compartment of the bin.

4. The plastics could also be segregated further based on their types, grades and colours.

5. The program for the microcontroller could be written in a way that the controller stores new voltage readings for opening the lid of the bin compartments if a setting button is pressed.

REFERENCES

[1] S. Gupta, K. Moban, R. K. Prasad, S. Gupta, and A. Kansal, "Solid waste management in indian: Options and opportunities in resource conversion", *Resour Conserv Recycl.,* vol. 24, no. 2, pp. 137-154, 1998.
     [http://dx.doi.org/10.1016/S0921-3449(98)00033-0]

[2] S. Sakai, S.E. Sawell, and A.J. Chandler, "World trends in municipal solid waste management", In: *Environmental Preservation Centre* Kyoto University: Japan. 16:341, 1996.

[3] S. Yamazaki, H. Nakane, and A. Tanaka, "Basic analysis of a metal detector", *IEEE Trans. Instrum. Meas.,* vol. 51, no. 4, pp. 810-814, 2002.
     [http://dx.doi.org/10.1109/TIM.2002.803397]

[4] M.H. Russel, M.H. Chowdhury, S.N. Uddin, A. Newaz, and M.M. Talukder, "Development of automatic smart waste sorter machine", *International Conference on Mechanical, Industrial and Materials Engineering,* 2013pp. 1-8

[5] S.J. Ojolo, J.I. Orisaleye, A.O. Adelaja, and O. Kilanko, "Design and development of waste sorting machine", *JETEAS,* vol. 2, no. 4, pp. 576-580, 2011.

[6] M. Rodionov, and T. Nakata, "Design of an optimal waste utilization system: A case study in St. Petersburg, Russia", *Sustainability,* vol. 3, no. 9, pp. 1486-1509, 2011.
     [http://dx.doi.org/10.3390/su3091486]

[7] V. Yadav, S. Karmakar, A.K. Dikshit, and S. Vanjari, "A feasibility study for the locations of waste transfer stations in urban centers: A case study on the city of nashik, india", *J. Clean. Prod.,* vol. 126, pp. 191-205, 2016.
     [http://dx.doi.org/10.1016/j.jclepro.2016.03.017]

[8] Y. Glouche, A. Sinha, and P. Couderc, "A Smart waste management with self-describing complex objects", *Int. J. Adv. Intell. Sys. IARIA,* vol. 8, no. 1 & 2, pp. 1-16, 2015.

[9] A. Sinha, and P. Couderc, "Smart bin for incompatible waste items in ICAS", *The 9th International Conference on Autonomic and Autonomous Systems,* pp. 40-45, 2013.

[10] M. Arebey, M. Hannan, H. Basri, R. Begum, and H. Abdullah, "Integrated technologies for solid waste bin monitoring system", In: *Environmental Monitoring and Assessment.* Springer Netherlands. 2011, pp. 399–408.
     [http://dx.doi.org/10.1007/s10661-010-1642-x]

[11] M. Hannan, M. Arebey, H. Basri, and R. Begum, "Rfid application in municipal solid waste management system", *Aust. J. Basic Appl. Sci.,* vol. 4, no. 10, pp. 5314-5319, 2010.

[12] B. Chowdhury, and M. Chowdhury, "Rfid-based real-time smart waste management system", *Proceedings of the Australasian Telecommunication Networks and Applications (ATNAC 2007),* pp. 175-180, 2007.
[http://dx.doi.org/10.1109/ATNAC.2007.4665232]

[13] S.K. Adzimah, and S. Anthony, "Design of garbage sorting machine", *Am. J. Eng. Appl. Sci.,* vol. 2, no. 2, pp. 428-437, 2009.
[http://dx.doi.org/10.3844/ajeassp.2009.428.437]

[14] F.R. Falayi, B.O. Adetuyi, and A. Adesina, "Development of small scale municipal waste sorter", *J. Eng. Appl. Sci.,* vol. 2007, no. 2, pp. 1640-1645, 2007.

[15] D. A. Wahab, A. Hussain, E. Scavino, M. M. Mustafa, and H. Basri, "Development of a prototype automated sorting system for plastic recycling", *Am. J. Appl. Sci.,* vol. 3, no. 7, pp. 1924-1928, 2006.
[http://dx.doi.org/10.3844/ajassp.2006.1924.1928]

CHAPTER 11

Development of a Fire Detection and Extinguishing Robot

Ilesanmi Afolabi Daniyan[1,*], Adefemi Adeodu[2], Bankole Oladapo[3], Vincent Balogun[4] and Ididiong Etudor[5]

[1] *Department of Industrial Engineering, Tshwane University of Technology, Pretoria, South Africa*

[2] *Department of Mechanical Engineering, University of South Africa, Florida, South Africa*

[3] *School of Engineering and Sustainable Development, De Montfort University Leicester, Leicester, UK*

[4] *Department of Mechanical Engineering, Edo State University, Iyamho, Nigeria*

[5] *Department of Mechanical Engineering, Afe Babalola University, Ado Ekiti, Nigeria*

Abstract: Robotics finds application in firefighting services. This work considers a robot that is able to detect fire and extinguish it. The robot operates automatically, avoiding obstacles, and at the same time, it is capable of detecting, tracking, and extinguishing flames. To achieve the best performance with an effective implementation, a modular design strategy was adopted, where the robot is divided into a number of logical modules based on functionality. The design consists of five main modules: the master controller, motor control, proximity control, fire detection and fire extinguishing module. Each module is associated with appropriate sensors and actuators. The information from various sensors and key hardware elements is processed *via* the PIC18F452 microcontroller. This is then interfaced with the master controller which coordinates and schedules the task of the entire robotic system. The performance evaluation indicates the robot's capability to detect and extinguish flames, hence, this work will generate interest as well as innovations in the field of robotics while working towards practical and obtainable solutions to save lives and mitigate the risk of property damage.

Keywords: Actuators, Microcontrollers, Modules, Sensors.

INTRODUCTION

In recent years, the advent of the Fourth Industrial Revolution (4IR) has continually promoted the capabilities of robotic solutions as an intelligent and autonomous system with effective service delivery [1]. Robotics has gained popularity due to the advancement of many technologies of computing and

* **Corresponding authors Ilesanmi Afolabi Daniyan:** Department of Industrial Engineering, Tshwane University of Technology, Pretoria 0001, South Africa; Tel: +27 (064) 5298778; E-mail: afolabiilesanmi@yahoo.com

nanotechnology, thus, making product development and services efficient, easier, and comfortable [2, 3]. Most robots have augmented microcontrollers (a small control system with input and output control capabilities). Robotic situation finds useful application in the dangerous or hazardous environment as well as difficult-to-reach area. Firefighting is an important but dangerous occupation and can cause the death of personnel, destruction of properties, and permanent disabilities [4, 5]. Hence, the associated risks can be mitigated with the deployment of a robot for firefighting operations. Firefighting must be able to provide a quick emergency response with the ability to get to the scene of a fire quickly and safely extinguish the fire in order to prevent further damage and reduce fatalities [6 - 8]. Since technology has finally bridged the gap between firefighting and machines, allowing for a more efficient and effective method of firefighting, a robot can be designed to function by itself or be controlled from a distance, which means that firefighting and rescue activities could be executed without putting the firefighters at risk through the use of robot technology. The robot employs a microcontroller designed to accept input from a set of electrical signals and give output in the form of electrical signals in response to commands programmed into the device. The microcontrollers, which interpret a human interface and send electrical signals to the rest of an electronic device, are often implemented as small, embedded processors found in many modern electronic devices. The versatility of microcontrollers is because they are considered tiny cost-effective computers which operate like a single integrated circuit. In addition, they are programmable, affordable, and lightweight, require minimum power, and range from different varieties to suit every need [9 - 11]. A firefighting mobile robot is one of the solutions that are able to reduce the hazards and risks of a firefighter [12 - 14]. Xu *et al.* [15] designed a firefighter robot capable of searching, detecting, and extinguishing fire in a small floor plan of a house. It also has the capability to move about and avoid obstacles. The navigation of the robot is achieved by the data provided by a line tracker and ultrasound transducers. The deployment of the extinguishing device is implemented with a custom arm controlled by servos. Ratnesh *et al.* [16] developed an approach toward the implementation of a firefighting robot. It employs the concept of environmental sensing and awareness as well as proportional motor control. The robot processes information from its sensors and hardware elements while ultraviolet, infrared, and visible light are used to detect the components of the environment. The robot is capable of fighting tunnel fire, industry fire, and military applications. Saravanan [17] developed an integrated semi-autonomous firefighting mobile robot. The system controls four D.C. motors powered by an Atmega2560 microcontroller and it is controlled autonomously by the navigation system. Furthermore, Poonam *et al.* [18] developed an intelligent fire extinguishing system that uses a smoke sensor, flame sensor, and temperature sensor for fire detection. The system detects the fire location and extinguishes it using the sprinklers method. Meanwhile, most researchers are working on feasible solutions to help robots think more efficiently, move and navigate in a smoother way [19-21]. This work considers the development of a fire detection, tracking, and extinguishing robot. The systems' capability includes the ability to operate automatically, avoid obstacles and at the same time, find and track and extinguish flames *via* a modular design strategy. The robot has the ability to act and think independently mimicking humans but with much degree of flexibility. The development of sensor networks and the maturity of robotics suggest that mobile agents can be employed for tasks that involve prompt perception and reaction to an external stimulus, even when the reaction involves a significant amount of mechanical actions. This provides the opportunity to pass on to robots, tasks that traditionally humans had to do but were inherently life-threatening such as fire-fighting, hence, fire-fighting is an obvious candidate for automation in this regard. Given the number of lives lost regularly during firefighting operations, the system can be deployed to mitigate the hazards and associated risks involved in the firefighting operation. Experience suggests that designing a fire-fighting system with sensors and robots is within the reach of the current sensor network and mobile agent technologies. Furthermore, the techniques developed in this work will carry over to other areas involving sensing and reacting to stimulus, where it is desired to replace humans with an automated mobile agent. Fire accidents can occur anywhere at any time and it rapidly spreads causing havoc.

METHODOLOGY

The design consists of five main modules: the master controller, motor control, proximity control, fire detection and fire extinguishing module. Each module is associated with appropriate sensors and actuators. The information from various sensors and key hardware elements is processed *via* the PIC18F452 microcontroller. This is then interfaced with the master controller which coordinates and schedules the task of the entire robotic system. The following specifications were made for the design of the fire-detecting and extinguishing robot: six (6) panels of sensors, in order to have a 360-degree view for flame detection, servo, and DC motors, ultrasonic sensors for obstacle detection, and CO_2 fire extinguisher.

The robot has a modular design as illustrated in Fig. (**1**), where the entire task to be performed by the robot is split into:

The flame detection and tracking module comprises two flame sensors (YL-38), based on the light wavelength ranging from 750-1200 nm for flame detection, and an ultrasonic sensor (HC-SR04) for obstacle detection and avoidance. The signals measured by the sensors in real-time serve as input into the microcontroller.

For the controller module, the threshold value for flame detection is preset on the microcontroller and the output from the sensors is received as input in the microcontroller. The signal is amplified and converted to a digital signal on the microcontroller using the Analog to Digital Converter (ADC). The digital signal is then processed and compared with the threshold. Once the threshold is exceeded, the microcontroller activates the DC motor which moves the robot for firefighting operation.

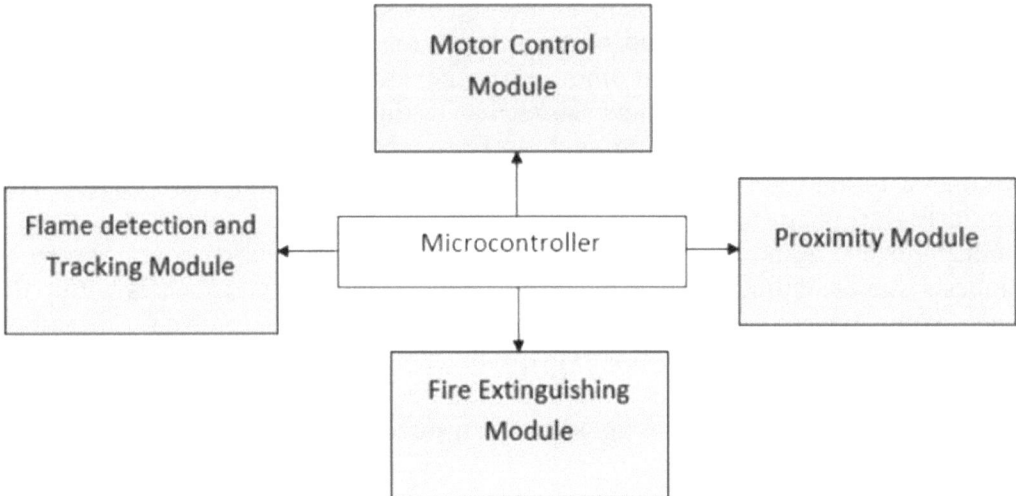

Fig. (1). Modular design of the robot.

The motor control module has a 12V DC motor which is activated by the microcontroller whenever there is a need for firefighting operation.

On detecting a flame source, the fire extinguishing module is activated and the robot moves in the direction of the flame detected and stops at a safe distance which ranges from 0.5-1.0 m depending on the flame intensity. The safe distance is to prevent the flame intensity from damaging the robot. The "stop" action is automatically followed by the activation of the fire extinguisher which is sprayed at the flame until the flame intensity falls below the threshold preset on the microcontroller as measured in real-time by the flame sensor.

Design Constraints and Specifications

Aside from the modular design strategy, some basic constraints were placed on the robot. These constraints served as a design guideline for the implementation of the system.

1. The robot must be able to identify and extinguish the flame in less than 5 minutes and return to its initial position in less than 2 minutes after extinguishing the flame.

2. The robot is designed not to exceed a size of 31 cm in width, 31 cm in length and 27 cm in height.

3. The battery must be able to last at least 30 minutes without the need for recharge.

4. The fire extinguisher system must work for 30 seconds without a refill.

5. The robot will not exceed 11.3 kg in weight.

Mechanical Design

The design was made using Autodesk Inventor Professional 2018. The chassis Setup was constructed using Actobotics aluminum. The choice of aluminum is because of its excellent mechanical properties such as an excellent lightweight-to-strength ratio, good formability and excellent corrosion resistance. Its lightweight-to-strength ratio implies that the robot will be highly sustainable in terms of cost economics, power requirement as well as environmental friendliness. Figs. (**2** and **3**) present the chassis model of the robot as well as the front view.

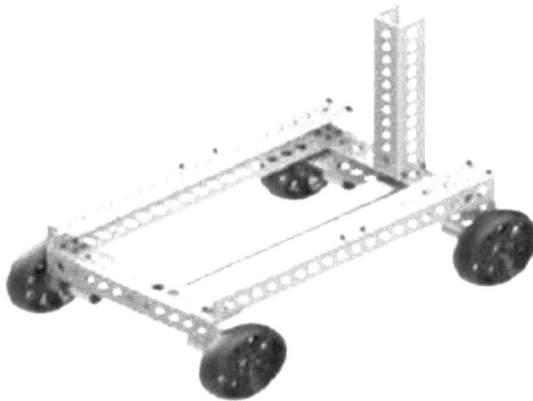

Fig. (2). The chassis model of the robot.

Fig. (3). The front view of the robot.

Motor Size Determination

Based on the estimate of the total robot's weight, a 12 V was found sufficient for the service and functional requirements.

The Circuit Integration and Analysis

The Proteus 8.0 was employed for the circuit design and from the schematic of the Proteus circuits for the respective modules, the control boards were made with all circuit components soldered on a stripboard. The connections made were tested for continuity to ensure that no components were connected in the wrong manner. Since the framework of the robot was made of aluminum material and could conduct current, the circuit boards were insulated with rubber standoffs to ensure that the boards were completely isolated from the framework to avoid short-circuiting of the system. Figs. (**4** and **5**) show the integrated circuit board for the motor and the controller.

Fig. (4). Motor control circuit board.

Fig. (5). Master control circuit board.

Fig. (**6**) shows the front view of the robot, after assembly and integration of all the necessary components.

Fig. (6). The developed fire-detecting and extinguishing robot (front view).

Performance Evaluation

The performance evaluation of the developed fire extinguishing robot was performed under different flames conditions namely; candle, kerosene stove, domestic gas cooker, bunsen burner, welding flame, furnace and petrol flame in order to evaluate the detection and extinguishing time of the robot. The results obtained are presented in Table **1**.

Table 1. The results obtained.

Flame source	Detection time (sec)	Distance from frame (m)	Extinguishing time (sec)
Candle	0	0.5	0
Kerosene stove	20	0.5	15
Domestic gas cooker	18	0.65	30
Bunsen burner	15	0.75	40

(Table 1) cont.....

Flame source	Detection time (sec)	Distance from frame (m)	Extinguishing time (sec)
Welding flame	15	0.75	45
Furnace	10	1.0	-
Petrol flame	5	1.0	90

RESULTS AND DISCUSSION

The robot demonstrated excellent navigating and obstacle avoidance abilities when approaching a flame source for extinguishing. Furthermore, in line with the design requirement, the robot could also keep a safe distance ranging between 0.5-1.0m depending on the intensity of the flame. The detection and the extinguishing time of the developed robot are presented in Table **1**.

From Fig. (**7**), it was observed that the detection time decreases with the intensity of the flame. The higher the flame intensity, the faster the detection time and response of the robot. This is due to the fact that the threshold value for the flame detection preset on the micro-controller forms the basis for the activation of the extinguishing module. On the other hand, the extinguishing time increases with the intensity of the flame. The intensity of the candle flame at a distance of 0.5 m was far less than the threshold value for the flame intensity, hence, there is no basis for the flame extinguishing. The fuel in the kerosene stove was made to overflow before the extinguisher could detect the presence of the flame after 20 seconds taking a total of 15 seconds for extinguishing the flame. The results indicate that the developed fire extinguishing robot has the capacity to detect flame sources early enough to forestall the destruction of lives and property and at a safe distance from the flame, it also possesses the capacity to extinguish the flame promptly.

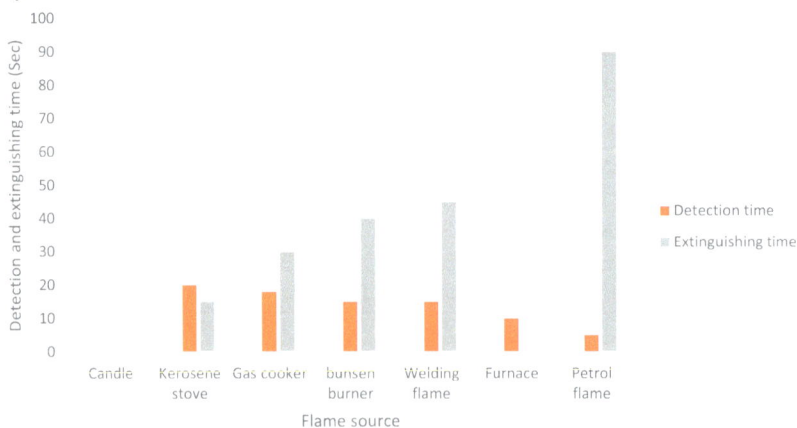

Fig. (7). The detection and the extinguishing time.

CONCLUSION

The development of a fire-extinguishing robot for firefighting was carried out. The work was motivated by the desire to develop a system that can detect fires and take appropriate actions, without any human intervention. The results obtained indicate that the robot can independently identify and extinguish the flame in less than 5 minutes and return to the initial position in less than 2 minutes after extinguishing the flame at a safe distance ranging from 1.5m-1.0m depending on the flame intensity. The robot can be used in educational institutes, malls, industries, work laces, or rather, anywhere. Its cost-effectiveness makes it viable. Its successful completion helps to generate interest as well as innovations in the field of robotics while working towards a practical and obtainable solution to save lives and mitigate the risk of property damage.

REFERENCES

[1]   R.C. Luo, and L. Su. Kuo, "Autonomous fire detection system for intelligent security robot", *IEEE/ASME Transactions on Mechatronics*, vol. 12, no. 3, pp. 274-281, 2007.
[http://dx.doi.org/10.1109/TMECH.2007.897260]

[2]   A. Sivas, and S. Kalaimani, "Wireless surveillance robot with motion detection and live video transmission", *Int. J. Engg. Res*, vol. 1, no. 6, pp. 14-22, 2013.

[3]   M.A. Mi-Yusouf, N.S. Sani, and A. Zainal, "Development of firefighting robot (QRob)", *Int. J. Adv. Comput. Sci. Appl.*, vol. 10, no. 1, pp. 142-147, 2019.

[4]   C. Sangaralingam, and R. Sathish, "Design of surveillance robot with obstacle sensing and movement control using arm controller", *Int. J. Engg. Res.*, vol. 2, no. 2, pp. 201-206, 2014.

[5]   T. Nandkishor, R.M.K. Satbhai, A.V. Patil, and M. Patil, "Fire fighting robot", *Int. J. Recent Innov. Trends Comput. Commun.*, vol. 4, no. 4, pp. 799-803, 2016.

[6]   J-H. Kim, S. Jo, and B.Y. Lattimer, "Feature selection for intelligent firefighting robot classification of fire, smoke, and thermal reflections using thermal infrared images", *J. Sens.*, vol. 2016, pp. 1-13, 2016.
[http://dx.doi.org/10.1155/2016/8410731]

[7]   A. Çakir, and N.F.E. Ezzulddın, "Fire-extinguishing robot design by using arduino", *J. Comput. Eng.*, vol. 18, no. 6, pp. 113-119, 2016.

[8]   J. Raju, S.S. Mohammed, J.V. Paul, G.A. John, and D.S. Nair, "Development and implementation of arduino microcontroller based dual mode fire extinguishing robot", *IEEE Int'l. Conference on Intelligent Techniques in Control, Optimization & Signal Processing INCOS*, 2017 pp. 1-5
[http://dx.doi.org/10.1109/ITCOSP.2017.8303141]

[9]   T.M. Khoon, P. Sebastian, and A.S. Saman, "Autonomous firefighting mobile platform. international symposium on robotics and intelligent sensors", *Procedia Eng.*, vol. 41, pp. 1145-1153, 2012.
[http://dx.doi.org/10.1016/j.proeng.2012.07.294]

[10]   C. F. Tan, S. M Liew, M. R. Alkahari, S. S. S. Ranjit, M. R. Said, W. Chen, G. W. M. Rauterberg, D. Sivakumar, and D. M. Sivarao, "Firefighting mobile robot: State of the art and recent development", *Aust. J. Basic Appl. Sci.*, vol. 7, no. 10, pp. 220-230, 2013.

[11]   M.H. Ali, S. Shamishev, and A. Aitmaganbayev, "Development of a network-based autonomous firefighting robot", *15th International Conference on Informatics in Control, Automation and Robotics*, 2018 pp. 525-532
[http://dx.doi.org/10.5220/0006928305250533]

[12]   D. William, G. Hector, B. Kevin, and D. Daisy, "An autonomous firefighting robot", In: *ECE at FIU*Miami, FL, 2010.

[13]   R.N. Haksar, and M. Schwager, "Distributed deep reinforcement learning for fighting forest fires with a network of aerial robots", *IEEE/RSJ Int'l Conference on Intelligent Robots and Systems IROS,* pp. 1067-1074, 2018 .
[http://dx.doi.org/10.1109/IROS.2018.8593539]

[14]   C. Xin, D. Qiao, S. Hongjie, L. Chunhe, and Z. Haikuan, "Design and implementation of debris search and rescue robot system based on internet of things", *International Conference on Smart Grid and Electrical Automation,* pp. 303-307, 2018 .
[http://dx.doi.org/10.1109/ICSGEA.2018.00082]

[15]   H. Xu, H. Chen, C. Cai, X. Guo, J. Fang, and Z. Sun, "Design and implementation of mobile robot remote fire alarm system", *Int'l Conference,* pp. 43-47, 2011 .
[http://dx.doi.org/10.1109/ISIE.2011.46]

[16]   M. Ratnesh, M. Kumbhare, and S.S. Kumbhlkar, "Firefighting robot: An approach", *Ind. Stre. Rese., II,* vol. 2012, no. 2, pp. 1-4, 2012.

[17]   P. Saravanan, "Design and development of integrated semi-autonomous firefighting mobile robot", *Int. J. of Eng. Sci. and Innov. Techn.,* vol. 4, no. 2, pp. 146-151, 2015.

[18]   S. Poonam, G. Rutika, P. Siddhi, and A. Kaldate, "Intelligent fire extinguisher system", *IOSR J. Comput. Eng.,* vol. XVI, no. 1, pp. 59-61, 2014.

[19]   Michie, D. Expert systems and robotics. New York: Wiley, 1985.

[20]   Barros, T. T. T and Lages, W. F. Development of a firefighting robot for educational competitions. RiE 2012, Prague.

[21]   T. AlHaza, A. Alsadoon, Z. Alhusinan, M. Jarwali, and K. Alsaif, "New concept for indoor firefighting robot", *Proc., Soc. and Beha. Sci.,* vol. 195, pp. 2343-2352, 2015.

CHAPTER 12

Performance Simulation of a Solar-Powered and Hand Gesture Controlled Lawn Robot using Dynamic Movement Primitives (DMP)

Adefemi Adeodu[1,*], Rendani Maladzhi[1], Mukondeleli Grace Kana-kana[2], Ilesanmi Afolabi Daniyan[2] and Kazeem Aderemi Bello[3]

[1] *Department of Mechanical Engineering, University of South Africa, Florida, South Africa*

[2] *Department of Industrial Engineering, Tshwane University of Technology, Pretoria 0001, South Africa*

[3] *Department of Mechanical Engineering, Federal University, Oye-Ekiti, Nigeria*

Abstract: Hand-gesture interpretation and control in robotics describe the interconnection between human and machine elements in the computer vision world. Pruning a structured environment is time-consuming and labor-intensive. Therefore, it requires management by a self-propelled machine. The path planning mode allows the robot to move along a specified path. Various studies on lawn mower robots focus more on obstacle avoidance with hand gesture interpretation and control implemented to take care of path definition. This study targets the development of a solar-powered lawn mower robot using hand gesture control as a path-planning technique. The robotic system continuously operates using charged batteries *via* solar energy stored in photovoltaic cells. The robot control mechanism was implemented *via* the use of infrared sensors to avoid obstruction on its path, and hand gesture interpretation *via* a DSP processor for path planning. The performance evaluation of the robot was based on field experiments and simulations using SolidWorks, defined in terms of area covered, lawn availability, energy utility, and optimum turning velocity. The evaluation revealed that the machine's efficiency is almost 100% based on the area covered, the percentage availability of the robot is 95%, and the average energy utility of 7.7 KWh was also obtained. The optimum turning velocity of 0.096 m/s at work with a completion time of 20 minutes was obtained by simulation. This robot is useful for any environment, both structured and semi-structured.

Keywords: Area covered, Digital signal processing, Hand gesture, Robotic, Obstacle avoidance.

* **Corresponding author Adefemi Adeodu:** Department of Mechanical Engineering, University of South Africa, Florida, South Africa; Tel: +27 (087) 223718; E-mail: eadeodao@unisa.ac.za

Ilesanmi Afolabi Daniyan (Ed.)

INTRODUCTION

Hand gesture recognition and interpretation in robotics is a type of machine-human interaction implemented in the field of computers when dealing with path planning. On this basis, robotic systems take definite instructions for carrying out tasks relative to directions. This work is an extension of the work originally presented in ICAST by Adeodu *et al.* [1]. The study of gesture identification and interpretation distinguishes defined body movements and communicates the message to the user to establish the link between humans and machines [2, 3].

Hand gesture control finds a wide range of applications in telerobotics, where machine systems are naturally manipulated with such telerobotic communication [4 - 6] to serve information relating to directions to the machine, such as left, right, *etc.* Its application in the lawn mower robot is a simple and unique method of controlling the geometric movement of the robot as a means of path planning. The aesthetic value maintenance of an environment, structured, or semi-structured is generally labor-intensive and time-consuming [1]. Therefore, the need for executing the task effectively is of utmost importance.

Several works have been reported on the development of lawn mowers. For instance, the CWRU is a grass cutter robot that captures and processes images using a 1.83 GHz run on the Windows XP operating system with 4GB memory. Efficient computations and robust obstacle identification methods were used in this robot based on image hue and intensity [7 - 10]. Deranda *et al.* [11] developed a smart robot lawn mower that does not require a boundary cable for its operation to reduce the cutting cycle time and operational cost. Ahamed and Ziauddin [12] developed a small-scale electric lawn mower while Tanimola *et al.* [13] developed a solar-powered lawn mower that can harness the potential of solar energy as a renewable source. In addition, Namoco *et al.* [14] developed a mechanical push lawn mower with a double cylinder for blade spinning to increase the system's capacity and operational efficiency while Moore *et al.* [15] performed the measurement and analysis of a lawn mower in terms of its performance and noise level. Many other works have been reported on the development of autonomous lawn mowers and systems that require the installation of boundary wire around the perimeter of the lawn area to keep the machine within the predetermined area and path to be cut [16 - 19].

The contribution of this work was demonstrated by configuring hand gestures to define the path of a solar-powered grass-cutting robot. This was achieved by showing hand movements in front of a high-resolution camera linked to a digital signal processing (DSP)-based embedded board, which captured the image in real-time. An algorithm based on a DSP processor TMS320DM642EVM was used to evaluate actuation *via* many hand gestures without contact with the surface of the screen. Real-time performance was made possible, which extended the limit of applications to include those with high frame rates. The robot's gesture learning was accomplished *via* dynamic movement primitives (DMP). Dynamic movement primitives are represented by dynamical systems and can be generalized by modifying initial values and tuning parameters based on environmental changes [20]. Hence, they can be executed from different initial and final poses with a similar but not the same trajectory, which eases the task of making robots more social. Therefore, this work approaches cognition from a dynamic system's perspective.

Dynamic Movement Primitives

Dynamic Movement Primitives (DMP) is a method for trajectory control planning that was originally derived from Stefan Schaal's laboratory in 2006 and later updated by Auke Ijspeert in 2013 [21]. A detailed and highly illustrative explanation of DMP can be found in [22]. The trajectory that defines the DMP is a point attractor dynamic equation, which is presented in Eqs. (1-11).

$$y = \alpha_y \left\{ \beta_y (g - y) - y \right\} \tag{1}$$

Where y is the system state, g is the goal, α and β are gain terms. This kind of system can be seen as a proportional derivative control signal governing the system to the target [23]. To modify the trajectory by adding a force term, Eq. (1) is transformed as follows:

$$y = \alpha_y \left\{ \beta_y (g - y) - y \right\} + f \tag{2}$$

Defining a non-linear function that achieves the desired behavior is not a trivial task. To generalize and make the trajectory time-independent, a new system called the canonical dynamic system is introduced, which has very simple dynamics:

$$X = -\alpha_x X \tag{3}$$

The canonical system originated at some arbitrary value such as xo = 1 and goes to infinity. Hence the forcing function f can be defined over a basic function $\ni_i$ defined on canonical system x.

$$f(x, g) = \frac{\sum_{I=1}^{N} \ni_i w_i}{\sum_{I=1}^{N} \ni_i} X (g - y_0) \tag{4}$$

Where y_o is the initial position of the system and wi is a weighting for the given basic function $\ni_i$. Usually, basis are defined like Gaussian functions centered at ci, where h_i is the variance.

$$\ni_i = \exp(-h_i(x - C_i)2) \tag{5}$$

Defining a non-linear function to achieve the desired behavior is not a trivial task. To generalize and make the trajectory time-independent, a new system called the canonical dynamic system is introduced, with very simple dynamics. The forcing function is a set of Gaussian functions that are activated as the canonical system x converges to its target [23]. Their weighted function is normalized and then multiplied by the X (g-y_o) term, which is both a diminishing and spatial scaling term. Incorporating the term x into the forcing function guarantees that the contribution of the forcing term goes to zero over time as the canonical system does. Moreover, spatial scaling means that once the system is set up to follow a defined trajectory to a specified goal, the goal can be adjusted further to get a scaled version of the trajectory. Finally, generalization in terms of temporal scaling is also sought to be able to follow the same trajectory at different speeds. Another term, τ, is added to the dynamic system to achieve flexibility.

$$y = \tau^2 \propto_y \{\beta_y(g - y) - y) + f\} \tag{6}$$

$$X = \tau(-\propto_x X) \tag{7}$$

In order to slow sown the system, τ is span from 0 to 1 and to speed it up, τ is set greater than 1. Having a forcing term to make the system conformed to a trajectory pathway as it converges to a target point. In order to condition the path of the trajectory of the system, the acceleration of the system is impacted using the forcing term. Therefore, from the desired trajectory $y_d = \{y_d(t_0) \ldots \ldots \ldots y_d(t_n)\}$. the forcing term is needed to generate the trajectory.

$$f_d = y_d - \propto_y \{\beta_y(g - y) - y\} \tag{8}$$

Since the forcing term is comprised of a weighted summation of the basis function which is activated through time [4], an optimization technique such as locally weighted regression (LWR) can be selected to calculate the weight w_i such that the forcing function matches the desired trajectory f_d. The objective function to be minimized is,

$$\sum_t \ni i(t)\{f_d(t) - w_i(x(t)(g - y_0)))2\} \tag{9}$$

And the solution is,

$$w_i = \frac{S^T \ni_i f_d}{S^T \ni_i S} \tag{10}$$

$$\text{Where S} = \begin{pmatrix} X_{to}(g - y_o) \\ X_{tN}(g - y_o) \end{pmatrix} \text{ and } \ni_i = \begin{pmatrix} \ni_i (t_o) \dots \dots \dots 0 \\ 0 \dots \dots \dots \dots \dots \dots 0 \\ 0 \; \dots \dots \dots \dots \ni_i (t_n) \end{pmatrix} \tag{11}$$

SYSTEM ARCHITECTURE AND DESCRIPTION

The robot uses captured frames and a live video stream within a fixed time range for gesture control, based on the concept of natural computing, which is accessible to anyone regardless of hand anthropometry [2, 3]. Gestures were classified using the principle of component analysis, implemented in embedded Matlab, Simulink, and Code Composer Studio (CCS). An XDS 560 PCI JTAG Emulator was fitted into a PCI slot to enhance high-speed RTDX on the enabled processor, allowing for data transfer at a rate of 2 Mb per second [2, 3]. Other features are the same as those applicable to the robot developed by Adeodu *et al.* [1]. Figs. (**1** and **2**) show the flow process diagram of the robot's hand gesture control and the circuit diagram, respectively.

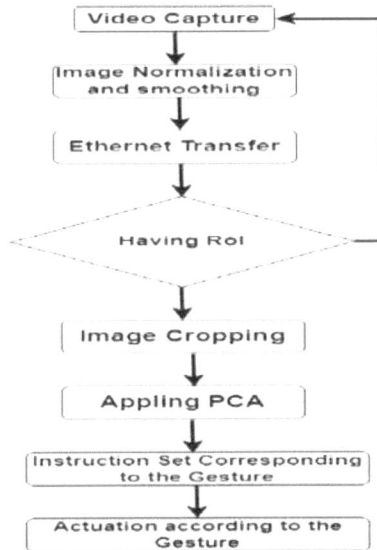

Fig. (1). Process flow diagram for robot hand gesture control [2].

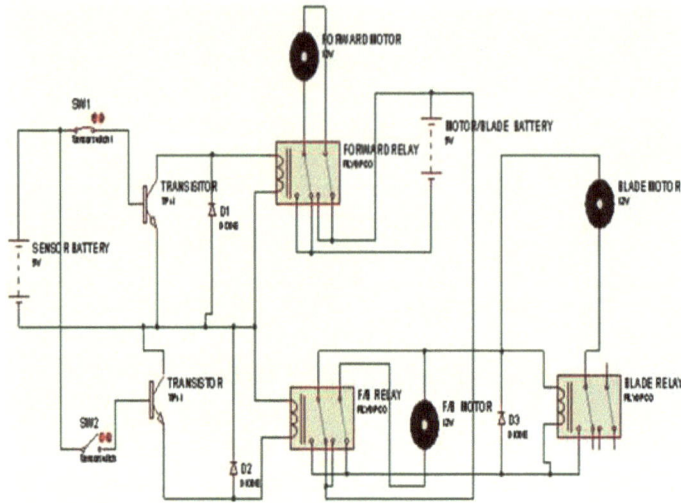

Fig. (2). Diagram of the circuit of the robot [1].

Table **1** presents the robot's components and specifications.

Table 1. Robot Components and specifications.

S/N	Component	Specification
1	Solar Panel	18 V (20 Watts)
2	DC Motor for cutter	3000 rpm (12 V, 700 mA)
3	Batteries	Lithium cell 12 V (10 Ah)
4	LCD Screen	16 by 2 and 5 by 7 pixel matrix
5	Microcontroller	ATmega 328 arduino Output PCB A-13400-3
6	High Resolution Camera	Model RF625 Range from 5 to 1400 mm Linearity in the range of ±0.1
7	Infrared Sensor	Model HC-SR04 Range of 2 to 400 cm Accuracy over 3 mm
8	Blade	Stainless steel Thickness is 0.5 mm
9	Wheel of the Robot, DC Motors and Gears	BLDC 300 rpm
10	DSP Processor	TMS320DM642EVM

SYSTEM DESIGN

Obstacle avoidance and hand gesture control modules are the main design architecture for the robot. The robot navigates along a defined path avoiding collision against any obstruction.

Design of Module for the Obstacle Avoidance

This is arranged such that colour and distance are detected *via* infrared sensors TCRT5000 using an echo resounding technique. The IR sensors emit waves that detect obstacles at their range when the echo of the reflected wave is received. The detailed module design and connection were presented by Adeodu *et al.* [1]. The flow chart of the module is shown in Fig. (**3**). Also, Figs. (**4** and **5**) present the adopted programming and circuit diagram respectively.

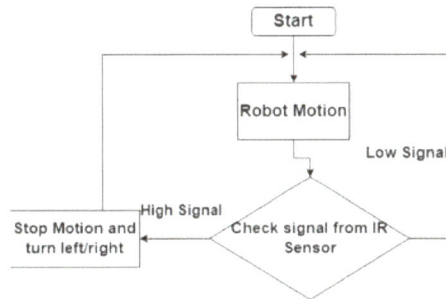

Fig. (3). Flow diagram module for the obstacle avoidance [1].

Fig. (4). Programming coding for the obstacle avoidance [1].

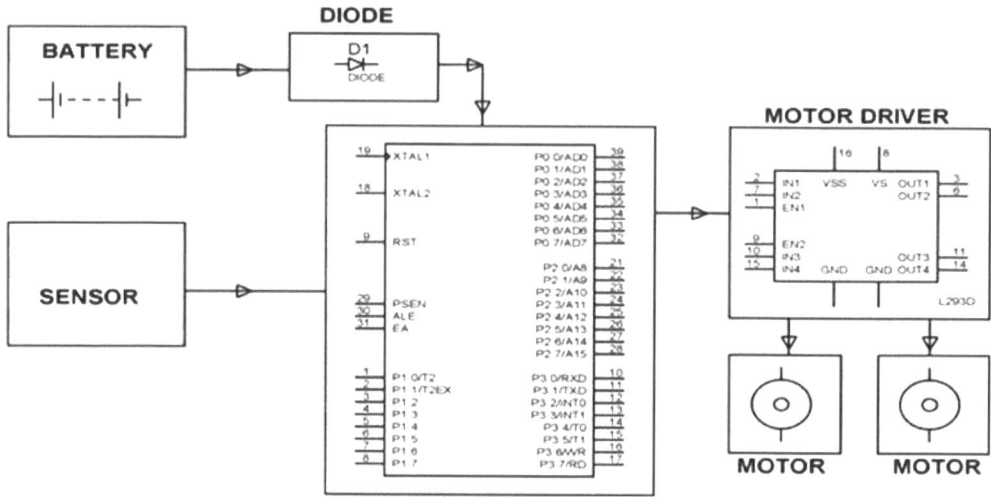

Fig. (5). Circuit Diagram of Module for the Obstacle Avoidance [1].

Module for the Hand Gesture Control

The design for hand gesture control for the robot was implemented in four stages *via* an embedded Matlab, Simulink, and code composer studio [2].

Gesture Extraction from Video Stream

Image capture and pre-processing were carried out frame by frame per one-fifth second. The image smoothing and normalization were executed within the DSP board as shown in Fig. (6). Additional information on smoothing and normalization is discussed by [24, 25]. A large reduction in the execution time for the algorithm was done by the board, after which the image frame was sent to the PC *via* UDP target to HOST Ethernet communication in Fig. (7). Fig. (8) shows the transfer of the data from the PC to the host [2].

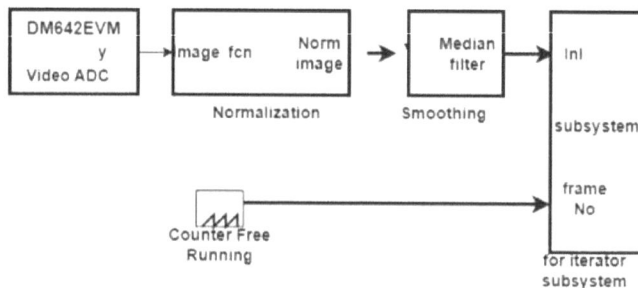

Fig. (6). Smoothing and normalization of DSP [3].

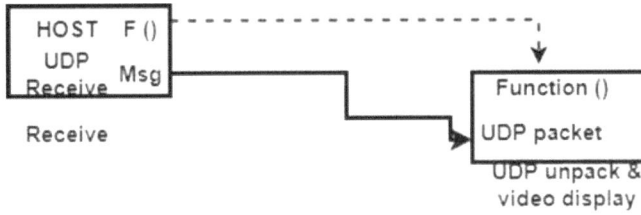

Fig. (7). UDP target to the host [3].

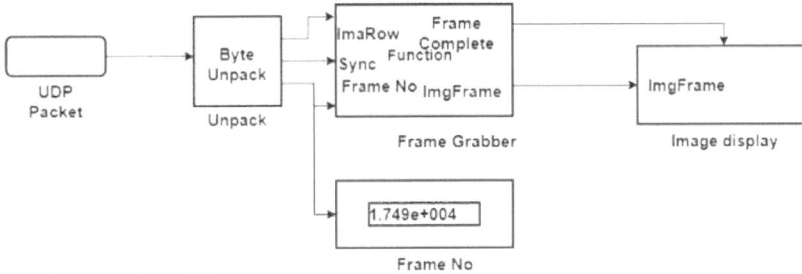

Fig. (8). Data transfer to the host [3].

Extraction of Region of Interest from the Frame

This aspect is normally referred to as Cropping, where the unwanted information from the selected frame is removed. The selected frame was changed to a binary image using global thresholding [2]. Fig. (**9**) shows the hand gesture frame prior to and afterward cropping.

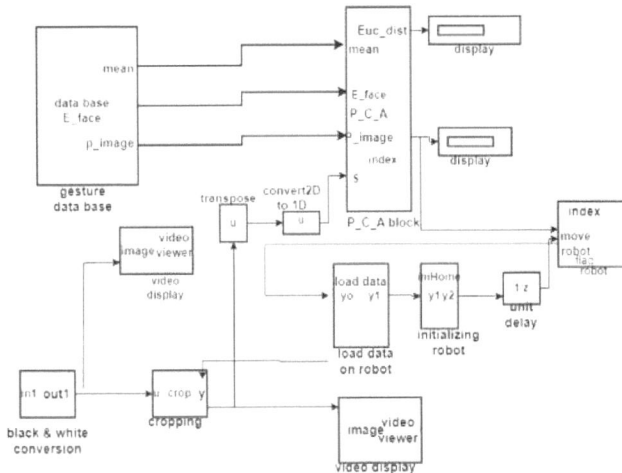

Fig. (9). Image after pre-processing [3].

Determination of Gesture and Pattern Matching

The component analysis (PCA) concept is preferable for pattern matching over other methods such as Artificial Neural Networks (ANN), which require more processing time [26]. It should be noted that cropped photographs were enlarged to correspond with data-based images.

Generation of Control Instruction for The Robot

The pattern for a specific motion is written down and saved in the database. Fig. (**10**) [3] shows how different hand gestures up to six were used to construct a specific move of the robot in precise angle and direction. The camera captures a gesture made in front of the robot. This is processed and compared to the records in the database. The action associated with the gesture was processed and delivered to the robot for execution. Fig. (**11**) depicts the processing of the hand gesture, robot movement, and view.

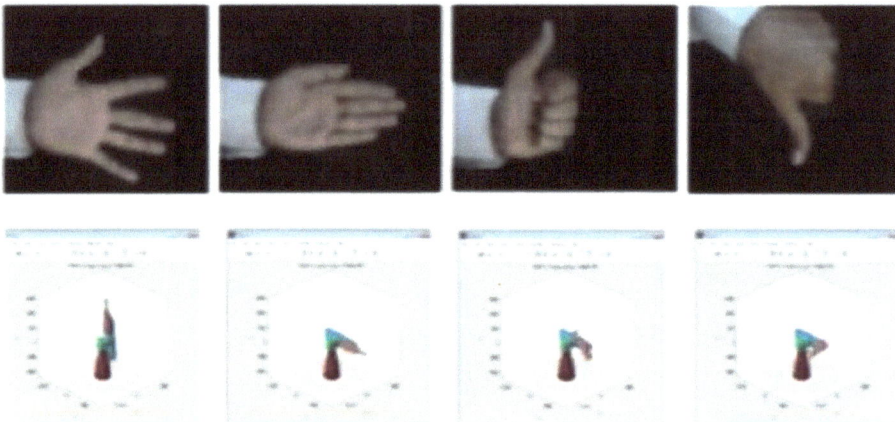

Fig. (10). Display of different hand gesture [3].

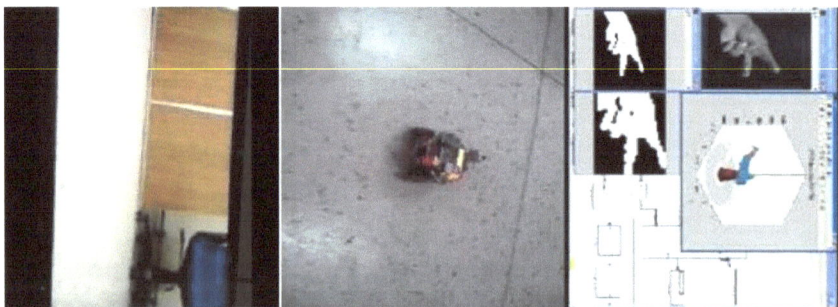

Fig. (11). Display of the processing of the hand gesture, movement of the robot and view.

Image Processing and Improvement

This is not a key sub-module of the hand gesture control; rather, it is located between the extraction of the region of interest and the determination of the gesture pattern. At this phase, the image quality is increased in terms of the pixel value by interacting between the AT Mega Arduino Uno microcontrollers and the Matlab GUI-based image editor. The microcontroller includes a boot loader that allows users to load fresh code [27]. Image processing is classified into three types: compression, enhancement and restoration, and measurement extraction [27]. Picture quality flaws are created by digitisation or setup flaws, which are rectified through picture enhancement. Once the system is in good working order, measurement extraction is used to obtain useful information such as color and tracking. The Matlab GUI for image enhancement is run on a computer with Matlab software and an Arduino Uno microcontroller board. This controls the robot by manipulating the motor driver circuit's action as the user controls the motor with hand gestures. The computer and Arduino communicate *via* a wireless system.

Image Colour Tracking Process

As input, the image movement frame with respect to color is used. This processing is critical to tracking, which employs Matlab coding [27]. Matlab tracks the color of the image movement direction (command) and delivers it straight to the Arduino Uno microcontroller *via* Arduino software connection. The Arduino, with the help of the motor driver circuit, controls the movement of the robot in the direction specified by the order. The Matlab program interprets the image color in the RGB format [2, 27] before processing it. To compensate for the camera's flipped photos, the RGB images are flipped in both rows and columns. The color on which the Matlab color tracking software is based is then extracted. The image also has dusty noise, which has been filtered using the median filter [27]. After that, the monochrome image is turned into black and white [27]. The area, centroid coordinates, and bounding box containing the color used for a specific command can be easily inferred from the image. The changes in x and y coordinates describe the hand's movement along the x and y axes, *i.e.*, right or left and upward or downward. The tracking software works as follows: The centroid and bounding box are tracked by the application. The program sends the direction command to the robot *via* software based on the movement of the centroid. It is taken into account that hand gesture movement is never absolute in one direction, however, before the command is sent to the robot, a random threshold of other coordinates is used to omit the changes [27].

Performance Simulation

SolidWorks 2018 Premium software was used to simulate the performance evaluation of the solar-powered hand-gesture-controlled lawn robot. This was utilized to investigate the variations in the robot's task completion time and centrifugal force as a function of turning velocity in order to establish the optimum turning velocity that will maximize cutting without losing stability and straying from the set path. Fig. (12) depicts the lawn model as well as its operational specifications.

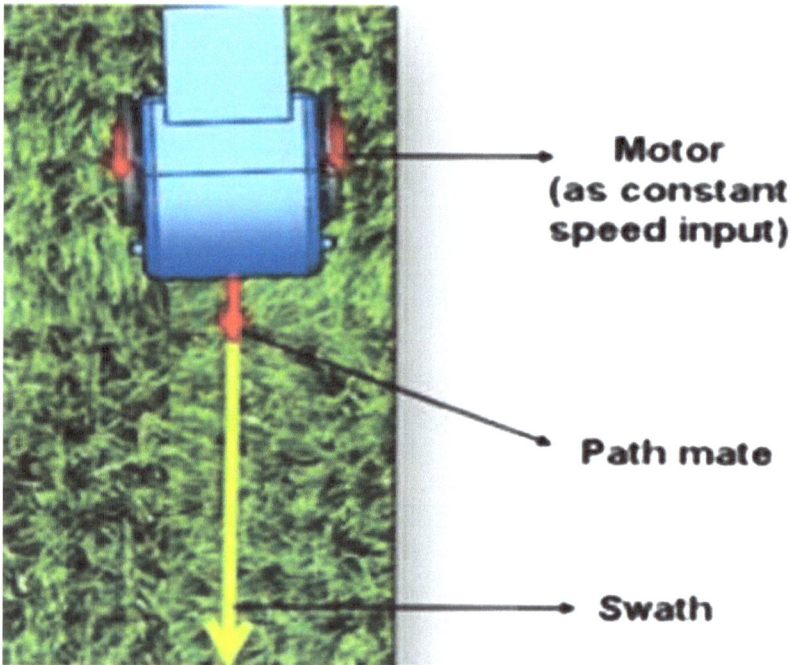

Fig. (12). Lawn model and operational details.

The task completion time was calculated by multiplying the entire distance travelled by the robot (area coverage) by the mowing speed. The total distance travelled was estimated by taking the perimeter of the traced rectangular path [28]. The blade length of the lawn robot is 0.55 m, according to the blade design. As a result, a cut ratio of 1 to 0.55 per square meter was assumed. As a result, the total area covered for the entire lawn was assessed to be 196 square meters. The simulation results are shown in Table **2**.

$$C_A = 196 x 0.55 = 107.8 m \tag{12}$$

The expression for work completion time is given as Eq. (13).

$$t_{wc} = \frac{107.8}{v \times 60}$$
(13)

Also, the expression for centrifugal force is given as Eq. (14).

$$F_c = \frac{mv^2}{r}$$
(14)

Where C_A = Covered Area (m)

t_{wc} is the work completion time (s)

v is the mowing velocity (m/s)

m is the mass of the robot (kg)

F_c is the centrifugal force (N)

r is the radius of curvature is taken as 0.2 m

Table 2. Results of the performance simulation.

S/N	Mowing Velocity (v) (m/s)	Work Completion Time (t_{wc}) (mins)	Centrifugal Force (F_c)(N)
1	0.0199	90.284757	0.079202
2	0.0398	45.142379	0.316808
3	0.0598	30.044593	0.715208
4	0.0797	22.542869	1.270418
5	0.0997	18.020729	1.988018
6	0.1196	15.022297	2.860832
7	0.1396	12.870105	3.897632
8	0.1595	11.264368	5.08805
9	0.1795	10.009285	6.44405
10	0.1994	9.0103644	7.952072

Discussion of Simulation Results

The findings of the lawn robot performance simulation are shown in Table **2**. Table **3** demonstrates that mowing velocity has an inverse relationship with task completion time, whereas mowing velocity has a direct relationship with the centrifugal force. The centrifugal force operating on the robot is roughly 8 N at the highest investigated mowing rate of 0.2 m/s, and a task completion duration of 9 minutes was attained. Fig. (**13**) depicts the ideal turning velocity curve that will minimize cutting time while not deviating from the designated direction by the gesture. The curve of task completion time and centrifugal force crossed at 0.096 m/s (optimum turning velocity), corresponding to a centrifugal force of 2 N, as shown in Fig. (**13**). This equates to a centrifugal force of 2 N and a job completion time of 20 minutes. This demonstrates that having a low centrifugal force acting on the robot makes it more stable for good cutting, which affects the cutting time [28 - 30].

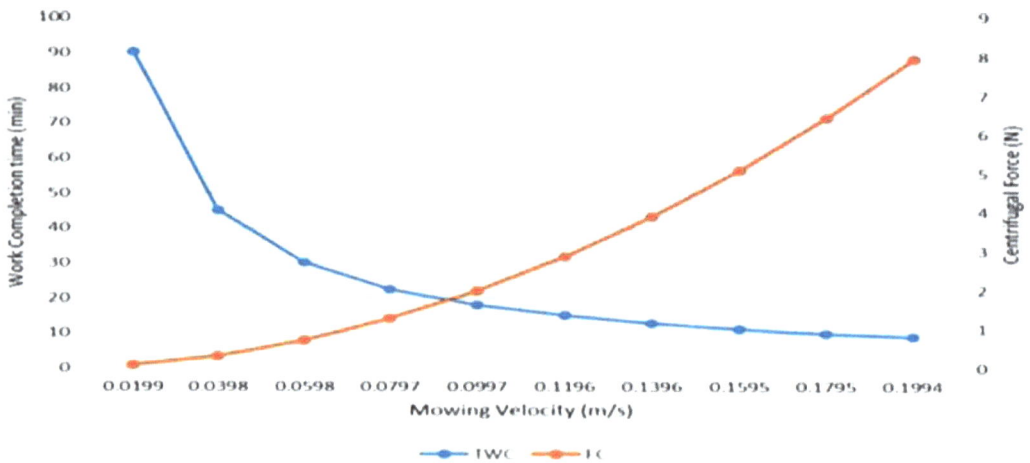

Fig. (13). Optimum turning velocity.

CONCLUSION

The study presents the creation of a solar-powered grass-cutting robot designed for a somewhat organized area. The installation of a solar-powered design and a hand gesture control module met the primary operational performance objectives of the maximum covered area, grass availability, and energy usage. The addition of hand gesture identification and control has substantially improved the robot's path planning and obstacle avoidance system while taking its kinematic constraint into account. Because the mowing velocity was properly managed, the work completion time was drastically reduced.

REFERENCES

[1]   A.O. Adeodu, I.A. Daniyan, T.S. Ebimoghan, and S.O. Akinola, "Development of an embeded obstacle avoidance and path planning autonomous solar grass cutting robot for semi-structured", *Environment.,* IEEE Xplore Digital Library, pp. 1-11, 2018.
        [http://dx.doi.org/10.1109/ICASTECH.2018.8506681]

[2]   J.L. Raheja, G.A. Rajsekhar, and A. Chaudhary, "Controlling a remotely located robot using hand gestures in real time: A dsp implementation", *5th IEEE International conference on Wireless Network and Embedded Systems,* India, Oct 16-17, pp. 1-5, 2016.
        [http://dx.doi.org/10.1109/WECON.2016.7993420]

[3]   J.L. Raheja, R. Shyam, G.A. Rajsekhar, and P.B. Prasad, "Real-time robotic control using hand gesture", In: *Robotic Systems - Applications, Control and Programming*, 2012.
        [http://dx.doi.org/10.5772/25512]

[4]   M. Moore, "A DSP-based real time image processing system", *In the Proceedings of the 6th International conference on signal processing applications and technology* Boston MA, 1995.

[5]   R. Verma, and A. Dev, "Vision based hand gesture recognition using finite state machines and fuzzy logic", *International Conference on Ultra-Modern Telecommunications & Workshops,* pp. 1-6, 2009.
        [http://dx.doi.org/10.1109/ICUMT.2009.5345425]

[6]   Y. Wu, and T.S. Huang, "Vision based gesture", *Proceedings of the International Gesture Workshop on Gesture-Based Communication in Human-Computer Interaction,* vol. 1739, pp. 103-115, 1999.

[7]   M. Kathleen Killough, and L.L. Crumpton, "An investigation of cumulative trauma disorders in the construction industry", *Int. J. Ind. Ergon.,* vol. 18, no. 5-6, pp. 399-405, 1996.
        [http://dx.doi.org/10.1016/0169-8141(95)00102-6]

[8]   J.M. Muggleton, R. Allen, and P.H. Chappell, "Hand and arm injuries associated with repetitive manual work in industry: a review of disorders, risk factors and preventive measures", *Ergonomics,* vol. 42, no. 5, pp. 714-739, 1999.
        [http://dx.doi.org/10.1080/001401399185405] [PMID: 10327893]

[9]   P. H. Batavia, S. A. Roth and S. Singh, "Autonomous coverage operations in semi-structured outdoor environments," International Conference on Intelligent Robots and Systems, Lausanne, Switzerland., vol. 1, pp. 743-749, 2002.

[10]   R. Keicher, and H. Seufert, "Automatic guidance for agricultural vehicles in europe", *Comput. Electron. Agric.,* vol. 25, no. 1-2, pp. 169-194, 2000.
        [http://dx.doi.org/10.1016/S0168-1699(99)00062-9]

[11]   J.M. Derander, P. Andersson, E. Wennerberg, A. Nitsche, E. Moen, and F. Labe, "Smart robot lawn mower", *Bachelor thesis Department of Computer Science and Engineering, Chalmers University of Technology, University of Gothenburg, Gothenburg, Sweden.,* pp. 1-70, 2018.

[12]   M.S. Ahamed, and A.T.M. Ziauddin, "Development of a small scale electric lawn mower", *Bangl. J. Agri. Engg.,* vol. 22, no. 1&2, pp. 37-44, 2011.

[13]   O.A. Tanimola, P.D. Diabana, and Y.O. Bankole, "Design and development of a solar powered lawn mower", *Int. J. Sci. Eng. Res.,* vol. 5, no. 6, pp. 215-220, 2014.

[14]   C.S. Namoco, J.D. Achas, R. Alcantara, J. Anora, and C.C. Buna, "Development of a mechanical push lawn mower with double-cylinder spinning blades", *Int'l. J. of Eng. Res. & Tech.,* vol. 2, no. 8, pp. 1861-1863, 2013.

[15]   M.D. Moore, "Measurement and analysis of lawn mower performance and noise", *Masters Thesis.* Iowa State University, USA. pp. 1-98, 1997.
        [http://dx.doi.org/10.31274/rtd-180813-8153]

[16]   A.G. Khan, and A. Qurishi, "Commercial grass cutting cum collecting machine", *IOSR J. Mech. Civ. Eng.,* vol. 10, no. 1, pp. 35-38, 2013.
        [http://dx.doi.org/10.9790/1684-1013538]

[17]   K. LaMott and S. Tosunoglu, Development of a robotic lawnmower as an adaptable platform for swappable remote and autonomous control packages. Florida Conference on Recent Advances in Robotics, FCRAR 2010 -Jacksonville, Florida, 2010.

[18]   S.H. Bhutada, and G.U. Shinde, "Design modification and performance comparison of lawn mower

machine by mulch and flat type cutting blade", *Int. J. Agric. Sci.,* vol. 9, no. 40, pp. 4638-4641, 2017.

[19]   I. Daniyan, V. Balogun, A. Adeodu, B. Oladapo, J.K. Peter, and K. Mpofu, "Development and performance evaluation of a robot for lawn mowing", *Proce. Manuf.,* vol. 49, pp. 42-48, 2020.
[http://dx.doi.org/10.1016/j.promfg.2020.06.009]

[20]   E. Rückert, and A. d'Avella, "Learned parametrized dynamic movement primitives with shared synergies for controlling robotic and musculoskeletal systems", *Front. Comput. Neurosci.,* vol. 7, p. 138, 2013.
[http://dx.doi.org/10.3389/fncom.2013.00138] [PMID: 24146647]

[21]   A.J. Ijspeert, J. Nakanishi, H. Hoffmann, P. Pastor, and S. Schaal, "Dynamical movement primitives: Learning attractor models for motor behaviors", *Neur. Comput.,* vol. 25, no. 2, pp. 328-373, 2013.
[http://dx.doi.org/10.1162/NECO_a_00393] [PMID: 23148415]

[22]   T. DeWolf, "Dynamic movement primitives' part 1: The basics", Available at: http://studywolf.word-press.com/2013/11/16/dynamic-movement-primitives-part-1-the-basics/

[23]   S. Pfeiffer, and C. Angulo, "Gesture learning and execution in a humanoid robot *via* dynamic movement primitives., *Patt. Recognit. Lett.,* vol. 67, pp. 100-107, 2015.
[http://dx.doi.org/10.1016/j.patrec.2015.07.042]

[24]   J.L. Raheja, B. Ajay, and A. Chaudhary, "Real time fabric defect detection system on an embedded DSP platform", *Optik,* vol. 124, no. 21, pp. 5280-5284, 2013.
[http://dx.doi.org/10.1016/j.ijleo.2013.03.038]

[25]   J.L. Raheja, S. Subramaniyam, and A. Chaudhary, "Real-time hand gesture recognition in FPGA", *Optik,* vol. 127, no. 20, pp. 9719-9726, 2016.
[http://dx.doi.org/10.1016/j.ijleo.2016.07.016]

[26]   S.R. Kota, J.L. Raheja, A. Gupta, A. Rathi, and S. Sharma, "Principal component analysis for gesture recognition using system", *Advances in Recent Technologies in Communication and Computing, ARTCom'09. International Conference on.,* pp. 732-737, 2009 .
[http://dx.doi.org/10.1109/ARTCom.2009.177]

[27]   S. Kar, A. Jana, D. Chatterjee, D. Mitra, S. Banerjee, D. Kundu, S. Ghosh, and S. Das, "Image processing based customized image editor and gesture controlled embedded robot coupled with voice control features", *Int. J. Adv. Comput. Sci. Appl.,* vol. 6, no. 11, pp. 91-96, 2015.
[http://dx.doi.org/10.14569/IJACSA.2015.061113]

[28]   S. Chauhan, "Motor torque calculations for electric vehicle", *Int. J. Sci. Technol. Res.,* vol. 4, no. 8, pp. 126-127, 2015.

[29]   S.G. Chouhan, S.A. Shaik, K.V. Krishnareedy, and S.R. Bandaru, "Design of a power autonomous solar powered lawn mower", *Int. J. Mech. Eng. Technol.,* vol. 8, no. 5, pp. 1-11, 2017.

[30]   "Lawn mower guide", Available at: http://www.lawnmowerguide.com/ (Accessed on: 31st May, 2022).

CHAPTER 13

Experimental Design and Modelling of Automated 4-Cylinder Engine Injector

Kazeem Aderemi Bello[1,*], **Abdulrahman Adama**[1], **Olasunkanmi Adekunle Odunaiya**[1] and **Cordelia Ochuole Omoyi**[2]

[1] *Department of Mechanical Engineering, Federal University, Oye Ekiti, Nigeria*

[2] *Department of Mechanical Engineering, University of Calabar, Calabar, Nigeria*

Abstract: The global emission regulations and fossil fuel pollution control have necessitated the need to study the effect of injector fuel splitting time and flow rate on engine performance. To achieve this, an automated prototyped 4-cylinder injector engine was developed to replicate the real-time activities of the injector system in the internal combustion (IC) engine. Arduino Nano open-source platform was used to integrate the various component parts such as the fuel injector, fuel tank, submersible fuel pump injector rail, transparent plastic chamber, flexible hose, Engine Control Unit (ECU), connecting wires, frame, Liquid Crystal Display (LCD), switch button, relay module, current sensor, potentiometer, Arduino nano, and pressure sensor that were used for the design experiment. Programmable circuit board microcontroller, Arduino (Integrated Development Environment) IDE, and C++ coding language were used to achieve the smart regulations of the injector operation system to replicate the real-time situation when the engine is running. This was achieved by incorporating Arduino microcontroller ATMEGA328, C++, and Arduino IDE software. The Arduino programming initiates the injection system and measures the injection output parameters. The system was designed to vary the splitting time delay between the four injectors and to measure the flow rate of the fuel injected. The experimental study showed that at a very high splitting time delay, the amount of fuel injected is more than the fuel injected at a relatively low splitting time delay with an average flow rate of 4.36 l/min at 50 microseconds and 0.02 l/min at 500 microseconds for high and low splitting time, respectively. This study will help the stakeholders in the automotive industry to virtualize the invisible situation of the fuel injector in real-time performance in the engine.

Keywords: Arduino Nano, Automation, Fuel injection, Microcontroller, Potentiometer.

* **Corresponding author Kazeem Aderemi Bello:** Department of Mechanical Engineering, Federal University, Oye Ekiti, Nigeria Tel: +234 (080) 36386760; E-mail: itanooluwaponmile@yahoo.com

Ilesanmi Afolabi Daniyan (Ed.)

INTRODUCTION

The fuel injection control system directly affects fuel efficiency and pollution level. Environment pollution and energy consumption have become serious concerns associated with engine control technology [1, 2]. The self-tuning control technique is applied to improve engine performance by controlling the engine speed and exhaust flow [3]. Most fuel injection systems are for gasoline or diesel applications. With the advent of electronic fuel injection (EFI), diesel and gasoline hardware have become similar. EFI's programmable firmware has permitted common hardware to be used with different fuels. Injector faults have negative effects on engine performance and they could cause engine misfiring, knocking, low thermal efficiency, or cause a total engine breakdown [4, 5]. During real-time monitoring of an engine injector performance, the Short-Term Fourier Transform (STFT) technique was used for fault diagnosis of fuel injection nozzles and knock detection [6, 7]. It is therefore necessary to prevent injectors' faults in an engine by monitoring their performance [8]. Condition monitoring of diesel engines can prevent unpredicted engine failures and the associated consequences [9]. An experimental study was carried out to instigate the use of gasoline fuel in a port injector in a single-cylinder Air-cooled HSDI Diesel Generator [6].The variable intake temperature and fuel split quantities were used to determine different combustion regime phases. This type of engine usually is available in four-stroke and two-stroke engines. In the four-stroke of ICE, each piston will undergo two strokes per crankshaft revolution in a certain order. The first order or stroke is the 'Induction' order where the intake valves are open and the exhaust valves are closed, and the piston moves downward increasing the volume of the combustion chamber thus allowing a mixture of air and fuel to enter the chamber. Next, is the 'Compression' stroke where both the intakes and exhaust valves will close as the piston moves upward until it reaches its top dead centre to decrease the volume of the chamber but increase pressure. Numerical simulations of compression ignition processes using a gasoline fuel injector at low pressure and hollow cone were investigated. The results showed good agreement with the experimental data in terms of pressure, thermal efficiency profile and emissions [7]. A fuel injection system (FIS) for a petrol engine is a system that utilizes fuel injectors instead of carburetors. The difference between carburetors and fuel injection is that fuel injection atomizes the fuel through a small nozzle under high pressure while carburetors rely on the suction power created by the air intake accelerated through a venturi tube to make the fuel into the airstream [10, 11]. The system works by determining the necessary amount of fuel, and its delivery into the engine is known as fuel metering [12, 13]. The early production of fuel metering was useful in mechanical methods and as time goes by, the modern system uses electronic metering which is more precise and smaller in size. At present time, more researchers are trying to convert the internal combustion system engine into a fuel injection system engine either a standard fuel injection system or direct fuel injection system because theoretically fuel injection systems bring more benefits to the consumers and are also environment friendly compared to the internal combustion system [14, 15]. The study on CFD modeling of the in-cylinder flow in direct-injection Diesel engines showed that the piston geometry had a minimum influence on the in-cylinder flow during the intake stroke and the first part of the compression stroke [16, 17]. However, the bowl shape plays a significant role in TDC and in the early stage of the expansion stroke by controlling both the ensemble-averaged mean and the turbulence velocity fields. The effects of the Abrasive Flow Machining (AFM) process on a direct injection (DI) Diesel engine fuel injector nozzle were studied [18, 19]. Geometry characterization techniques were developed to measure the microscopic variations inside the nozzle before and after the process [20]. Several studies had been conducted on empirical correlations of the nozzle geometry change due to the AFM process [21, 22] In modern automotive internal combustion engines, a variety of injection systems existed. A fuel injection system is designed and calibrated specifically for the types of fuel it will handle. Most fuel injection systems are for gasoline or diesel applications. With the advent of electronic fuel injection (EFI), diesel and gasoline hardware have become similar. EFI's programmable firmware has permitted common hardware to be used with different fuels. Basic components in a fuel injection system are a fuel injector, high-speed camera and electronic control unit (ECU) such as an injector driver and digital delay generator for the signal line while other components include a fuel tank, a fuel filter, a high-pressure pump and a pressure regulator for the fuel line. In the laboratory experiment, the high-pressure chamber is used as the main instrument to identify spray patterns. Some of the experiments use high-speed cameras with personal computers and ECU. The data gained will be shown on the personal computer automatically. The fuel injection types used in newer cars include four basic types; Single-point or throttle body injection, Port or multipoint fuel injection, Sequential fuel injection, and Direct injection. This study will serve as a teaching guide to stakeholders in the automotive industry to evaluate and improve the existing engine injector.

EXPERIMENTAL SET-UP

The experimental study of the automated cylinder injector system was achieved by using the following materials:

1. Fuel injector: Fuel injectors are nozzles that inject a spray of fuel into the air, basically the fuel injector used in this project is an Audi electronically controlled Injector system.

2. Fuel tank.

3. Submersible fuel pump: Submersible fuel-pump unit, consisting of an asynchronous electric motor and a multistage centrifugal pump of 40psi with a horsepower of 600hp (447.4kw) combined by a common shaft and a casing, and an electric motor.

4. Injector rail: A fuel rail is high-pressure tubing which takes fuel to the injectors in an internal combustion engine.

5. Transparent plastic chamber.

6. Automotive battery: An automotive battery or car battery is a rechargeable battery that is used to start a motor. Its main purpose is to provide an electric current to the electric-powered starting motor, which in turn starts the chemically-powered internal combustion engine that propels the vehicle.

7. Flexible hose.

8. Engine Control Unit (ECU).

9. Connecting wires.

10. Frame.

11. A Liquid Crystal Display (LCD) is a flat-panel display or another electronically modulated optical device that uses the light-modulating properties of liquid crystals combined with polarizers.

12. Push buttons A, B, and C.

13. Switch the button.

14. Relay module: a power relay module is an electrical switch that is operated by an electromagnet held in place by a spring, the armature leaves a gap in the magnetic circuit when the relay is de-energized.

15. Current sensor: Current sensors, also commonly referred to as current transformers or CTs, are devices that measure the current running through a wire by using the magnetic field to detect the current and generate a proportional output.

16. Potentiometer: a potentiometer is a three-terminal resistor with a sliding or a rotating contact that forms an adjustable voltage divider. If only two terminals are used, at one end and at the wiper, it acts as a variable resistor or rheostat.

17. Pressure sensor: A pressure sensor usually acts as a transducer and it generates a signal as a function of the pressure imposed. For this study, such a signal is electrical.

18. Arduino Nano: Arduino consists of both a physical programmable circuit board (often referred to as a microcontroller and a piece of software or IDE that runs on the computer Fig. (**1**). The programming language used in writing the coding is C++ language and Arduino IDE which is also known as the Arduino platform.

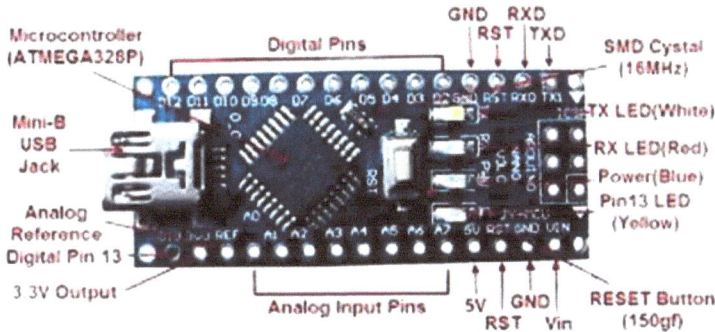

Fig. (1). Arduino Nano.

METHODS

Arduino Nano is an open-source platform used to integrate the material components in this study. It consists of a physical programmable circuit board microcontroller and IDE (Integrated Development Environment) that runs on the computer. C++ coding language and Arduino IDE are used to achieve the smart regulation of the injector.

The Arduino microcontroller with a specification number ATMEGA328 was programmed using C++ language with Arduino IDE software. The Arduino programming initiates the injection system and measures the injection output parameters [23]. Fig. (**2**) presents the Arduino IDE configuration scripts.

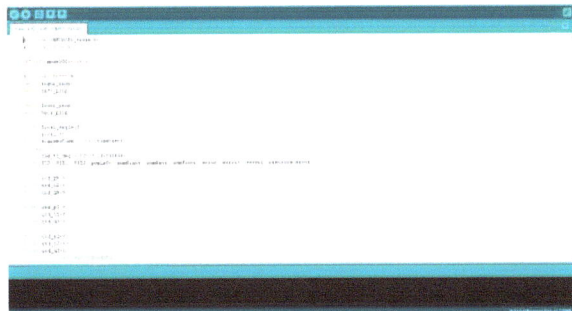

Fig. (2). Arduino IDE configuration Scripts.

THE MODELLING OF THE WORKING PRINCIPLE OF A 4-CYLINDER INJECTOR ENGINE

The injection system is designed to replicate the system of the 4-Cylinder injector engine [24]. The system is powered by a 12 V, 45 Ah capacity battery. The injection testing operation was programmed to commence operation by pressing the control board switch ON button. The control board has three buttons namely A, B, and C. The negative wire on the injectors is connected to the relay module which controls the "ON" and "OFF" injector switch buttons. The battery voltage is connected to a potentiometer to vary the voltage supplied to the injector. The system was computed to allow injection when the coming voltage supply is 7 volts upward. The current drawn varies according to the voltage supplied. The voltage, current, fuel splitting time, flow rate, and pressure are displayed on the LCD screen on the control board once the switch "ON" control button is pressed.

RESULT AND DISCUSSION

The injector system is designed to give detailed information on activities that take place during the fuel injection experimental modeling system. The LCD screen will display the information on fuel injection in real-time. The system is designed to depict the behaviour of an injector under different voltage regulations and the quantity of fuel split [25]. The splitting timing of the system was designed to increase from 500 microseconds to 50 microseconds. Table 1 shows the information obtained from the test and measurement under different voltages using three different injectors' performance regimes at different supplied voltages. The result obtained from time variation under different regimes is similar to the result of varying voltage supplied [26].

Fig. (3) presents the AutoCAD drawing of the experimental study.

Table 1. Flow rate versus voltage.

Volt. (v)	Flow rate (ml/min) Regime 1	Flow rate (ml/min) Regime 2	Flow rate (ml/min) Regime 3
7.0	0.00	0.00	0.00
8.0	0.00	0.00	0.00
9.0	0.00	0.00	0.00
10.0	14.23	14.64	14.45
10.5	23.34	22.93	23.47
11.0	33.65	33.46	33.95
11.5	40.12	40.36	40.94

(Table 1) cont.....

Volt. (v)	Flow rate (ml/min) Regime 1	Flow rate (ml/min) Regime 2	Flow rate (ml/min) Regime 3
12.0	45.87	46.23	46.61

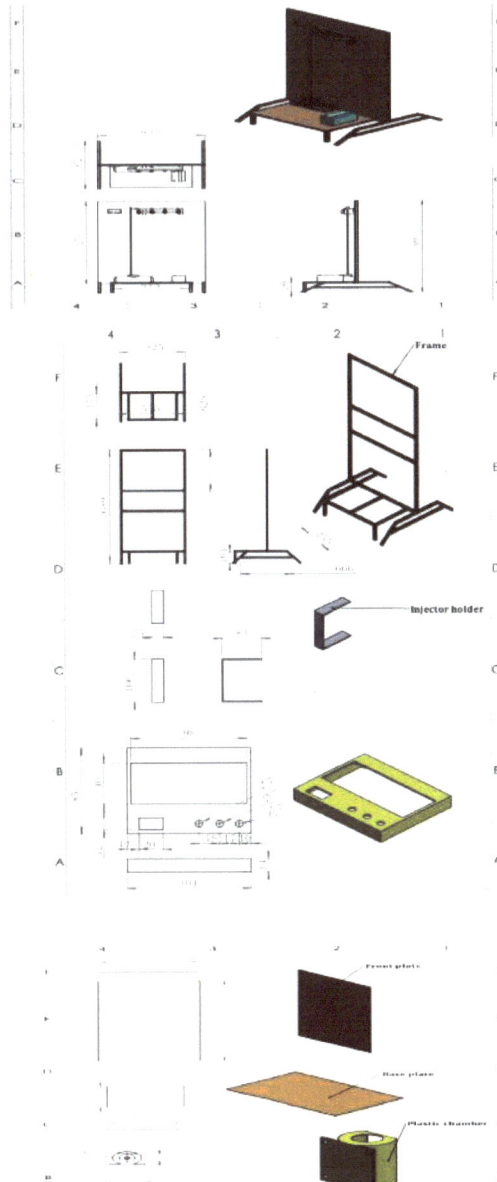

Fig. (3). Arduino IDE configuration Scripts.

It was observed that no fuel flowed at 7v, 8v, and 9v supply. It was also observed that the injector split at lower 7, 8, and 9 V voltage if the current is very high. It was noted that the injector spitted fuel at 10 volts was supplied. The current supplied to the injector was 4(A) because the potentiometer used could take more than 4(A) and with a lower current it couldnot have the injector splitting. It can also be observed from table 1 that with an increase in voltage, there will be high splits time, and flow rate. Fig. (**4**) depicts the fuel rate in the 3 regimes and the corresponding voltages. The trend is similar in each of the regimes, the higher the voltage supplied, the higher the flow rate. The graph indicates that at 35 volts, the corresponding flow rate starts to diminish. This is an indication that the optimality of the voltage supplied can be determined [27].

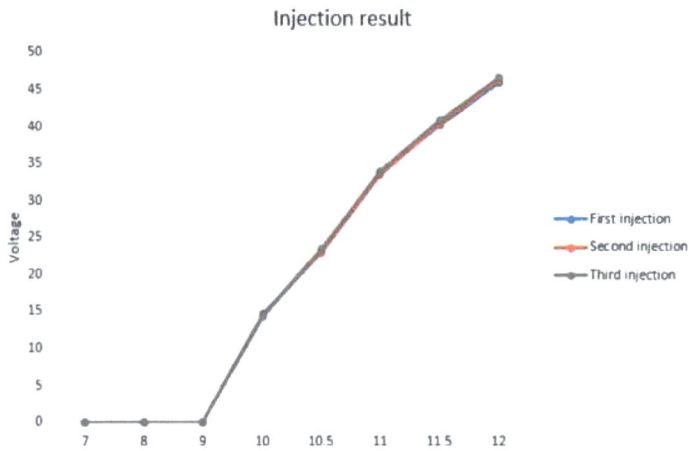

Fig. (4). Graph of Voltage supply versus flow rate.

Table **2** and Fig. (**5**) depict that splitting time delay is correlated to flow rate at five different experimental stages (1st, 2nd, 3rd, 4th, and 5th observable experiments). The ability to automate the regulation of splitting time can be used to improve the thermal efficiency of the engine. It can also be used by automotive manufacturers to build vehicles that can guide the drivers on speed selection depending on the drivers' experienced and road terrain [28].

Table 2. Splitting time and Flow rate regimes for five experimental tests.

Time delay (s)	1st (l/min)	2nd (l/min)	3rd (l/min)	4th (l/min)	5th (l/min)
50	4.33	4.41	4.35	4.40	4.34
140	0.12	0.16	0.14	0.17	0.13
320	0.07	0.08	0.07	0.07	0.09

(Table 2) cont.....

Time delay (s)	1st (l/min)	2nd (l/min)	3rd (l/min)	4th (l/min)	5th (l/min)
410	0.05	0.05	0.04	0.05	0.04
500	0.03	0.02	0.03	0.03	0.02

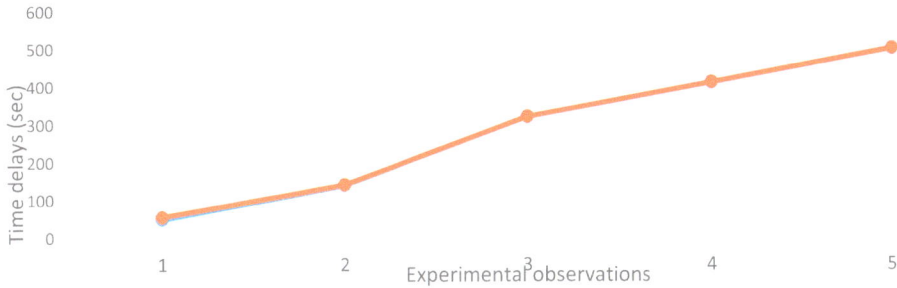

Fig. (5). Splitting times versus time delays.

Fig. (**6**) depicts the trend of flow rate in litre/minute at different splitting times. The flow rate is proportional to the unit of time. The lower the splitting time, the higher the flow rate. From fig. (**6**), it was noted that at the splitting time of 50 microseconds, the flow rate values were higher for the five samples with the flow rate values of 4.33, 4.41, 4.35, 4.40, and 4.34 l/min respectively for the five tested regimes. The lower the splitting time, the lower the value of the flow rate. This is an indication that fuel consumption is a function of acceleration and speed.

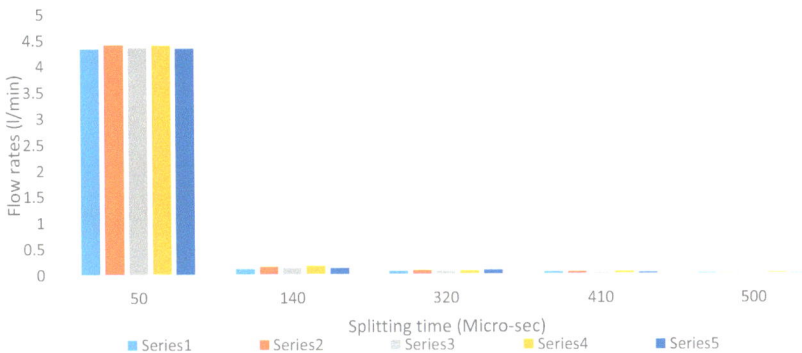

Fig. (6). Splitting time and flow rate results.

CONCLUSION

The experimental study of 4-cylinder injector testing has been successfully carried out with help of Arduino Nano configurations with other essential parts. The fuel splitting time and flow rate are adjusted experimentally until the desired optimization of fuel flow in the injector was achieved. An electronic fuel injection system provides more precise control of the fuel injection compared to a carburetor thus improving the efficiency of the engine with less fuel emission. A fuel injection system will make better fuel atomization and also better flow of air-fuel mixture into the piston. To get more accurate data, the test rig and setup need to be revised by adding more sensors like air intake, temperature sensor, and cylinder sensor.

REFERENCES

[1]   S. Shirvani, S. Shirvani, A.H. Shamekhi, R. Reitz, and F. Salehi, "Meeting euro6 emission regulations by multi-objective optimization of the injection strategy of two direct injectors in a DDFS engine", *Energy,* vol. 229, p. 120737, 2021.
[http://dx.doi.org/10.1016/j.energy.2021.120737]

[2]   B. Ashok, K.M. Usman, R. Vignesh, and U.A. Umar, "Model-based injector control map development to improve crdi engine performance and emissions for eucalyptus biofuel", *Energy,* vol. 246, p. 123355, 2022.
[http://dx.doi.org/10.1016/j.energy.2022.123355]

[3]   K. Mohiuddin, H. Kwon, M. Choi, and S. Park, "Experimental investigation on the effect of injector hole number on engine performance and particle number emissions in a light-duty diesel engine", *Int. J. Engine Res.,* vol. 22, no. 8, pp. 2689-2708, 2021.
[http://dx.doi.org/10.1177/1468087420934605]

[4]   A.D. Coutinho, "Fault injector for autonomous quadrotors",

[5]   Y. Cao, H. Liu, C. Song, Y. Jia, L. Yao, and J. Zhang, "Research on a fault test system based on fault injection", In: *Journal of Physics.* Conference Series, 2021, p. 012023.
[http://dx.doi.org/10.1088/1742-6596/1894/1/012023]

[6]   A. Taghizadeh-Alisaraei, and A. Mahdavian, "Fault detection of injectors in diesel engines using vibration time-frequency analysis", *Appl. Acoust.,* vol. 143, pp. 48-58, 2019.
[http://dx.doi.org/10.1016/j.apacoust.2018.09.002]

[7]   X. Yang, Q. Dong, J. Song, and T. Zhou, "Investigation of a method for online measurement of injection rate for a high-pressure common rail diesel engine injector under multiple-injection strategies", *Meas. Sci. Technol.,* vol. 33, no. 2, p. 025301, 2022.
[http://dx.doi.org/10.1088/1361-6501/ac3548]

[8]   Ç. Karatuğ, and Y. Arslanoğlu, "Importance of early fault diagnosis for marine diesel engines: a case study on efficiency management and environment", *Ships Offshore Struct.,* vol. 17, no. 2, pp. 472-480, 2022.
[http://dx.doi.org/10.1080/17445302.2020.1835077]

[9]   A. Golovan, I. Honcharuk, O. Deli, O. Kostenko, and Y. Nykyforov, "System of water vehicle power plant remote condition monitoring", In: *IOP Conference Series.* Materials Science and Engineering. p. 012049, 2021.
[http://dx.doi.org/10.1088/1757-899X/1199/1/012049]

[10]   A. Katijan, M.F.A. Latif, Q.F. Zahmani, S. Zaman, K.A. Kadir, and I. Veza, "An experimental study for emission of four stroke carbureted and fuel injection motorcycle engine", *J. Adv. Res. Fluid Mech.,* vol. 62, pp. 256-264, 2019.

[11]   N. Sutarna, I.N.L. Antara, and D.S. Anakottapary, "Fuel Consumption Analysis of Injection System and Carburetor System on Honda Beat Fi 2013", In: *Logic: Jurnal Rancang Bangun dan Teknologi.* vol. 62, pp. 256-264, 2019.

[http://dx.doi.org/10.31940/logic.v20i3.2170]

[12]   A. Ferrari, P. Pizzo, and R. Vitali, "Development and validation procedure of a 1D predictive model for simulation of a common rail fuel injection system controlled with a fuel metering valve", *SAE Int. J. Engines,* vol. 11, no. 4, pp. 401-422, 2018.
[http://dx.doi.org/10.4271/03-11-04-0027]

[13]   Y. Lu, Z. Zuo, C. Zhao, F. Zhang, and M. Du, "Study on dynamic characteristics and control algorithm design for fuel metering valve of high-pressure pump", *IFAC-PapersOnLine,* vol. 51, no. 31, pp. 930-935, 2018.
[http://dx.doi.org/10.1016/j.ifacol.2018.10.061]

[14]   Y. Huang, N.C. Surawski, Y. Zhuang, J.L. Zhou, and G. Hong, "Dual injection: An effective and efficient technology to use renewable fuels in spark ignition engines", *Renew. Sustain. Energy Rev.,* vol. 143, p. 110921, 2021.
[http://dx.doi.org/10.1016/j.rser.2021.110921]

[15]   Z. Li, Y. Wang, Z. Yin, H. Geng, R. Zhu, and X. Zhen, "Effect of injection strategy on a diesel/methanol dual-fuel direct-injection engine", *Appl. Therm. Eng.,* vol. 189, p. 116691, 2021.
[http://dx.doi.org/10.1016/j.applthermaleng.2021.116691]

[16]   S.K. Gugulothu, N.P. Kishore, V.P. Babu, and G. Sapre, "CFD analysis on different piston bowl geometries by using split injection techniques", *Acta Mech Malaysia,* vol. 2, no. 1, pp. 23-28, 2019.
[http://dx.doi.org/10.26480/amm.01.2019.23.28]

[17]   N. Seelam, S.K. Gugulothu, B. Bhasker, S. Mulugundum, and G.R. Sastry, "Investigating the role of fuel injection pressure and piston bowl geometries to enhance performance and emission characteristics of hydrogen-enriched diesel/1-pentanol fueled in crdi diesel engine", *Environ. Sci. Pollut. Res. Int.,* pp. 1-15, 2022.
[http://dx.doi.org/10.1007/s11356-021-18076-z] [PMID: 35028837]

[18]   M. Fang, T. Yu, and F. Xi, "An experimental investigation of abrasive suspension flow machining of injector nozzle based on orthogonal test design", *Int. J. Adv. Manuf. Technol.,* vol. 110, no. 3-4, pp. 1071-1082, 2020.
[http://dx.doi.org/10.1007/s00170-020-05914-6]

[19]   S. Han, F. Salvatore, J. Rech, J. Bajolet, and J. Courbon, "Effect of abrasive flow machining (AFM) finish of selective laser melting (SLM) internal channels on fatigue performance", *J. Manuf. Process.,* vol. 59, pp. 248-257, 2020.
[http://dx.doi.org/10.1016/j.jmapro.2020.09.065]

[20]   W. Zhu, Q. Ma, Z. Song, J. Lin, M. Li, and B. Li, "The effect of injection pressure on the microscopic migration characteristics by CO_2 flooding in heavy oil reservoirs", *Energy Sour. Recov. Util. Environ. Effec.,* vol. 44, no. 1, pp. 1459-1467, 2022.
[http://dx.doi.org/10.1080/15567036.2019.1644399]

[21]   Z. Liu, Y. Zhang, J. Fu, and J. Liu, "Multidimensional computational fluid dynamics combustion process modeling of a 6V150 diesel engine", *J. Therm. Sci. Eng. Appl.,* vol. 14, no. 10, p. 101009, 2022.
[http://dx.doi.org/10.1115/1.4054164]

[22]   S. Frommater, J. Neumann, and C. Hasse, "A phenomenological modelling framework for particle emission simulation in a direct-injection gasoline engine", *Int. J. Engine Res.,* vol. 22, no. 4, pp. 1166-1179, 2021.
[http://dx.doi.org/10.1177/1468087419895161]

[23]   Z. Zhang, H. Zhu, H. Guler, and Y. Shen, "Improved premixing in-line injection system for variable-rate orchard sprayers with arduino platform", *Comput. Electron. Agric.,* vol. 162, pp. 389-396, 2019.
[http://dx.doi.org/10.1016/j.compag.2019.04.023]

[24]   J. Kargul, M. Stuhldreher, D. Barba, C. Schenk, S. Bohac, and J. McDonald, "Benchmarking a 2018 toyota camry 2.5-liter atkinson cycle engine with cooled-EGR", *SAE Int. J. Adv. Curr. Prac. Mobi.,*

vol. 1, p. 601, 2019.
[http://dx.doi.org/10.4271/2019-01-0249]

[25] W.C. Tsai, and T.S. Zhan, "An experimental characterization for injection quantity of a high-pressure injector in GDI engines", *J. Low Power Electron. Appl.,* vol. 8, no. 4, p. 36, 2018.
[http://dx.doi.org/10.3390/jlpea8040036]

[26] V. Ayhan, and Y.M. Ece, "New application to reduce NOx emissions of diesel engines: Electronically controlled direct water injection at compression stroke", *Appl. Energy,* vol. 260, p. 114328, 2020.
[http://dx.doi.org/10.1016/j.apenergy.2019.114328]

[27] W.C. Tsai, "Optimization of operating parameters for stable and high operating performance of a GDI fuel injector system", *Energies,* vol. 13, no. 10, p. 2405, 2020.
[http://dx.doi.org/10.3390/en13102405]

[28] J. Finneran, C.P. Garner, M. Bassett, and J. Hall, "A review of split-cycle engines", *Int. J. Engine Res.,* vol. 21, no. 6, pp. 897-914, 2020.
[http://dx.doi.org/10.1177/1468087418789528]

SUBJECT INDEX

A

Abrasive flow machining (AFM) 189
AFM process 189
Air 27, 48
   conditioning system 27
   pressure control 48
Algorithms 8, 9, 97, 103, 172, 178
   clustering 8
   mathematical 8
Aluminum 152, 154, 166
   and iron of high torque 152, 154
   material 166
Applications 102, 104
   of industrial automation 102
   robotic 104
Arduino 96, 112, 117, 118, 187, 191
   board 112, 117, 118
   development boards 96
   programming 187, 191
Artificial intelligence 12, 103
   algorithms 103
   functions 12
Artificial neural networks (ANN) 3, 9, 99,
   100, 106, 180
Automated 1, 2, 5, 6, 13, 14, 20, 21, 27, 28,
   29, 50, 53, 55, 97, 101, 163
   guided vehicles (AGV) 28
   material handling systems 29
   mobile agent 163
   processes 2, 20
   production/assembly 14
   systems 1, 2, 5, 6, 13, 21, 27, 50, 53, 55,
   97, 101
   teller machine (ATM) 21, 101
Automatic 6, 21, 28
   assembly machines 28
   inspection system 21
   systems 6
Automation 1, 2, 6, 11, 13, 20, 21, 22, 27, 29,
   50, 55, 58
   cognitive 2, 50, 55, 58

environment 29
software 11
systems 2, 6, 20, 21, 22, 27, 50
technologies 1, 13

B

Biodiesel plant 108, 113, 129

C

CAD software 99
Calibrated flow rate sensors 112
CAM software 99
Canonical system 173, 174
Capacitive proximity sensor 111, 112, 152,
   153, 154, 156
CFD modeling 189
Chemical processes 28
Circuit, open-loop system 23
Cloud 10, 55, 111
   computing 10
   storage 55, 111
CNC machines 10, 106
Code composer studio (CCS) 156, 175, 178
Communication 20, 54, 57, 134, 136, 172
   microSD 54
   remote access 54
   system, wireless 136
   technologies 20, 57, 134
   telerobotic 172
Computer 58, 102, 105
   integrated manufacturing (CIM) 58
   numeric control systems 105
   systems, automating 102
Computer-aided 29, 57, 99
   design 29, 57, 99
   manufacturing 99
Computer numerical 22, 68, 102, 105
   control (CNC) 22, 68, 105
   control machines 102
Control 28, 56, 136

www.ingramcontent.com/pod-product-compliance
Lightning Source LLC
Chambersburg PA
CBHW050840220326
41598CB00006B/418